41 Topics in Current Chemistry
Fortschritte der chemischen Forschung

New Concepts I

Springer-Verlag
Berlin Heidelberg GmbH 1973

This series presents critical reviews of the present position and future trends in modern chemical research. It is addressed to all research and industrial chemists who wish to keep abreast of advances in their subject.

As a rule, contributions are specially commissioned. The editors and publishers will, however, always be pleased to receive suggestions and supplementary information. Papers are accepted for "Topics in Current Chemistry" in either German or English.

Any volume of the series may be purchased separately.

ISBN 978-3-662-15992-7   ISBN 978-3-540-46940-7 (eBook)
DOI 10.1007/978-3-540-46940-7

Originally published by Springer-Verlag Berlin Heidelberg New York in 1973
Softcover reprint of the hardcover 1st edition 1973
Library of Congress Catalog Card Number 51-5497.

Contents

Sulfuranes in Organic Reactions and Synthesis

Prof. Barry M. Trost

Department of Chemistry, University of Wisconsin, Madison, Wisconsin, USA

Contents

I. Introduction

Sulfuranes are a class of compounds in which sulfur expanded its formal valence shell from eight to ten electrons. There are two types of sulfuranes to be considered — π sulfuranes and σ sulfuranes. A π sulfurane possesses three sigma bonds and one pi bond. Only to the extent that back orbital overlap

$$\begin{array}{c}>\!\!\overset{\ominus}{C}-\overset{\oplus}{S}\!\!< \quad \longleftrightarrow \quad >\!\!C\!\!=\!\!S\!\!<: \qquad \pi \text{ sulfurane}\end{array}$$

between a filled *p* orbital on carbon and the empty *d* orbitals on sulfur is important has sulfur experienced valence shell expansion. These species are more commonly known as ylides. *A sigma sulfurane possesses four sigma bonds to sulfur in addition to the lone pair.* Such a valence shell

$$>\!\!\overset{|}{\underset{|}{S}}\!\!-\!\!: \qquad \sigma \text{ sulfurane}$$

expansion creates a strong driving force for reaction by allowing sulfur to return to the more stable octet state by loss of two electrons. For the π sulfuranes, such a reaction may be achieved with minimum loss of bonding energy by cleavage of the weak π bond generating a carbene and a dialkyl-sulfide (Eq. 1). Alternatively, the σ sulfuranes may return to the octet state

$$>\!\!C\!\!=\!\!S\!\!<: \quad \longrightarrow \quad <\!\!C: \ + \ :S\!\!<: \tag{1}$$

by either cleavage of one substituent with both bonding electrons to generate a carbanion and a sulfonium salt (Eq. 2) or of two substituents with concomitant bonding to generate a new carbon-carbon bond and a dialkyl-sulfide (Eq. 3).

$$\begin{array}{c}\overset{C}{\underset{C}{\overset{}{\underset{}{\overset{|}{S}}}}}\!\!-\!\!: \end{array} \quad \begin{cases} \longrightarrow \ >\!\!\overset{\oplus}{S}: \ + \ \overset{\ominus}{:}\!\!\overset{|}{C}\cdots \tag{2} \\[2em] \longrightarrow \ C\!\!-\!\!C+ \ :S\!\!\overset{C}{\underset{C}{<}}: \tag{3} \end{cases}$$

The two types of sulfuranes are related by consideration of their origin. Treatment of sulfonium salts with strong bases (normally organolithium) serves as the major technique of generating π sulfuranes. Nevertheless, it must be considered that sulfonium salts are potentially ambident in their behavior towards nucleophiles such as the organolithium bases. Instead of deprotonation, addition to the electrophilic sulfur may occur with formation of the sigma sulfurane. Indeed, many by-products observed in π sulfurane

reactions can now be attributable to the decomposition of σ sulfuranes. In some instances, σ sulfurane generation may compete so favorably that it becomes the predominant or exclusive mode of reaction. Both types of sulfuranes have applications in organic synthesis.

II. π Sulfurane Addition Reactions

The π sulfuranes are isoelectronic with diazo compounds and therefore may be expected to possess very similar chemical reactivity.[1-3] However, the zwitterionic character of these intermediates is greater than that of diazo compounds thereby facilitating their nucleophilic additions to carbonyl groups and Michael acceptors such as α,β-unsaturated carbonyl systems. Diphenylsulfoniumallylide serves as a typical example.[4] Thus, it reacts with 4-t-butylcyclohexanone to give the epoxide with predominantly the *trans* configuration and with methyl acrylate to yield the cyclopropane after 1,3-elimination. Many new types and new reactions of π

sulfuranes (sulfur ylides) remain to be discovered. Diphenylsulfoniumcyclopropylide is one such example.[5] The precursor sulfonium salt is readily available in large quantities by the route outlined below. The ylide is available from the salt irreversibly by treatment with dimsylsodium at -30 to $-40\,°C$ or preferably reversibly by treatment with powdered potassium hydroxide.

A. Spiroannelation

Reactions of diphenylsulfoniumcyclopropylide with aldehydes and ketones produces the very reactive oxaspiropentanes in high yield.[5,6] As the examples in Table 1 illustrate, electronically and sterically unreactive carbonyl groups produce the desired products well. The fascinating chemistry of these reactive compounds remains to be explored on the whole. However, they undergo facile rearrangement to cyclobutanones in quantitative yields upon acid treatment. Since rearrangement can be achieved upon

Table 1. Spiroannelations

Carbonyl	Oxaspiropentane	Cyclobutanone	% Yield
	a)		86
			94
			97

Table 1 (continued)

Carbonyl	Oxaspiropentane	Cyclobutanone	% Yield
PhCHO			87
			92
	a)		90
			59
			88
			53
			80
O‖PhCPh	a)		75

a) Not isolated but directly rearranged to cyclobutanone.
b) The yields have not been optimized in most cases.

working-up the initial reaction mixture, this procedure serves as an extremely general and versatile one-step cyclobutanone synthesis. The acid catalyzed epoxide rearrangement proceeds *via* carbonium ion intermeditates. Thus,

	2	3
M = H	83	17
Li	91	9

the oxaspiropentane from norbornanone *(1)* produces a mixture of two cyclobutanones *2* and *3* whose ratio depends upon the nature of the acid catalysis. With protonic acids, like fluoroboric or hydrochloric acids, a lower stereospecificity is observed than with Lewis acids like lithium perchlorate.

The greater selectivity with lithium salts may be attributed to the faster rate of cyclopropyl bond migration compared to rotation about the cyclopropyl carbon-norbornyl carbon bond in the carbonium ion intermediate *4a* when greater electron density is on oxygen. Since this reaction anneals a ring onto a carbonyl group with the production of spiro fused rings when the carbonyl partner is a cyclic ketone, this process represents a single member of a class of reactions termed spiroannelation. Johnson and coworkers independently examined the reactions of dimethylaminophenyloxosulfonium-cyclopropylide and reported a similar reaction in low yield.[7]

B. γ-Butyrolactone Synthesis

The ready availability of cyclobutanones by this one step very high yield process led to investigations of the synthetic applications of such intermediates. Oxidation to butyrolactones occurs readily and in yields of 70—90% utilizing either hypochlorous acid[8,9] or preferably basic hydroperoxide.[9,10]

$$ \qquad (4) $$

The reaction has the same characteristics as a Baeyer-Villager in that the more substituted carbon preferentially migrates (see Eq. 4) and migration occurs with retention of configuration at the migrating carbon (see Eq. 5). Thus, the overall synthetic event is to convert any carbonyl partner

$$ \qquad (5) $$

into a 5-mono or 5,5-disubstituted-γ-butyrolactone.

C. Geminal Alkylation[11]

Cyclobutanones are very readily substituted at their alpha positions *via* the corresponding enolates or enols. Thus, dibromination proceeds quantitatively by the dropwise addition of two moles of molecular bromine in

7

carbon tetrachloride at room temperature. Methanolic sodium methoxide effects ring cleavage at room temperature. The net result of this process involves introduction of two functionally different alkyl groups onto a carbonyl carbon — thus the terminology geminal alkylation. Methyl 2,2-

pentamethylene-4,4-dibromobutyric acid *(5)*, the product from cyclohexanone, illustrates some of the versatility of this approach. Silver ion

promoted methanolysis effects replacement of the bromides by methoxides with formation of an acetal. Acid hydrolysis frees the aldehyde functional group. This latter compound, 2,2-pentamethylenesuccinaldehydic acid methyl ester, was obtained in 86% overall yield from cyclohexanone. Obviously either the aldehyde or the ester may be further functionalized to generate a broad array of structural units. It must be noted that the aldehyde readily decarbonylates using Wilkinson's catalyst in refluxing acetonitrile.[12] Thus, this method even serves to convert a carbonyl group to a α-methyl carboxylic ester.

D. Secoalkylation[13]

The cyclobutanones derived from α,β-epoxyketones present several unique opportunities for further synthetic transformations. The oxaspiropentanes are formed in over 90% yield by the cyclopropylide reaction under the same conditions utilized previously. Rearrangement of the spiroepoxide in the presence of a second epoxide can be carried out with aqueous fluoroboric

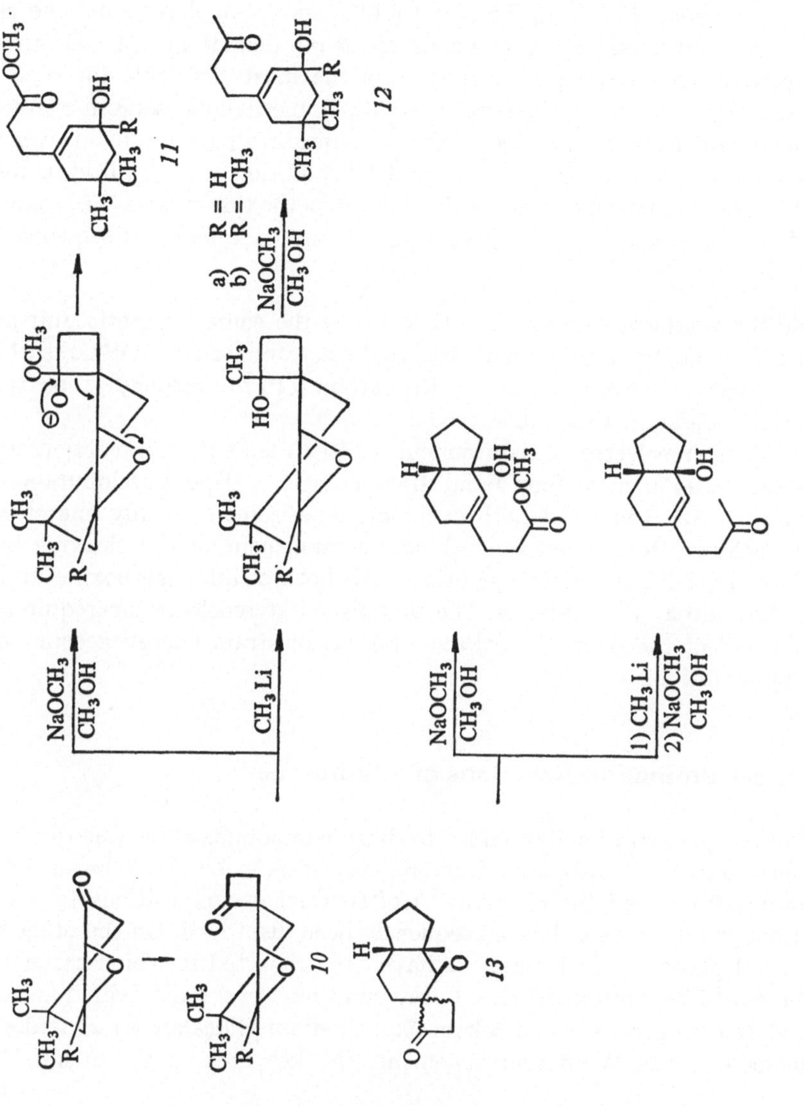

Scheme 1. Secoalkylation

acid in a two-phase system when the second epoxide is disubstituted, e.g. 9, R = H. When this epoxide is trisubstituted, e.g. 9, R = CH₃, oxalic acid in acetonitrile is to be preferred. Addition of a nucleophile to the cyclobutane carbonyl creates a good leaving group three carbon atoms removed from a good electron source, a negative oxygen.[14,15] In the case of the cyclobutanone *10a* from **5,5-dimethyl-2,3-epoxycyclohexanone**, the cleaving cyclobutyl and epoxide bonds are *trans* diaxial in the conformation depicted. This arrangement represents the most favorable for such four bond fragmentations. Dissolution of *10a* in methanolic sodium methoxide effects smooth fragmentation to the γ, δ-unsaturated ester *11* in over 90% yield from epoxycyclohexanone. Similarly, addition of methyllithium followed by fragmentation in methanolic sodium methoxide creates the 3-oxobutyl side chain as shown in compound *12*. The net result of this sequence is to

$$\overset{\ominus}{C}-C-\overset{\overset{\displaystyle O}{\|}}{C}$$

add the synthetic equivalent of $\overset{\ominus}{C}-C-\overset{O}{C}$ — the same synthetic unit added in a Michael reaction but with inverted electronic sense.[16] Wheras, the Michael acceptor is electrophilic at the carbon β to the carbonyl, this synthon is nucleophilic at this same carbon.

The stereoelectronic requirements of the fragmentation reaction are not as severe as in most four bond fragmentations. Epoxycyclobutanone *13* is an 87 : 13 mixture of both cyclobutanone isomers — only one of which can achieve the desired *trans* diaxial arrangement of the cleaving bonds. However, both fragment smoothly with both sodium methoxide and methyllithium as the addends. The diminished stereoelectronic requirements may reflect the over 55 kcal/mole release of strain energy accompanying fragmentation.

III. α-Elimination Reactions of π Sulfuranes

The comparison of sulfur ylides to diazo compounds raises the question of α-elimination to carbenes. Early reports of Franzen [17], Johnson [18], and Swain [19] claimed the observation of thermal decompositions to carbenes although one of these has subsequently been disproved. On the other hand, several photochemical reactions have been reported in which carbenes are the most likely intermediates. Corey and Chaykovsky discovered an Arndt-Eistert type process when β-keto oxysulfonium ylides are irradiated.[20] By analogy to the Wolff rearrangement, the key step in the Arndt-Eistert

$$(8)$$

sequence, invocation of a carbene intermediate appears most likely (see Eq. 8). Witkop and coworkers [21] observed a C—H insertion reaction upon irradiation of the stabilized ylide *16*. Such an insertion is characteristic of carbene chemistry.

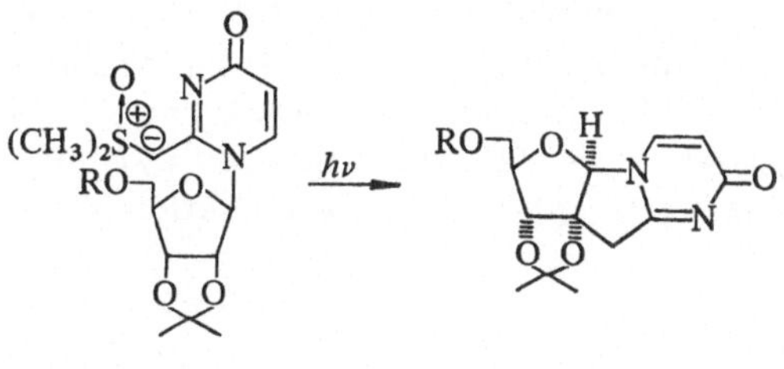

16

In sulfonium ylides, two examples have documented that they too are capable of undergoing α elimination to carbenes under photolytic, but not thermal, conditions. Dimethylsulfoniumphenacylide, *17*, yields a mixture of 7-benzoylnorcarane and acetophenone (by hydrogen abstraction from

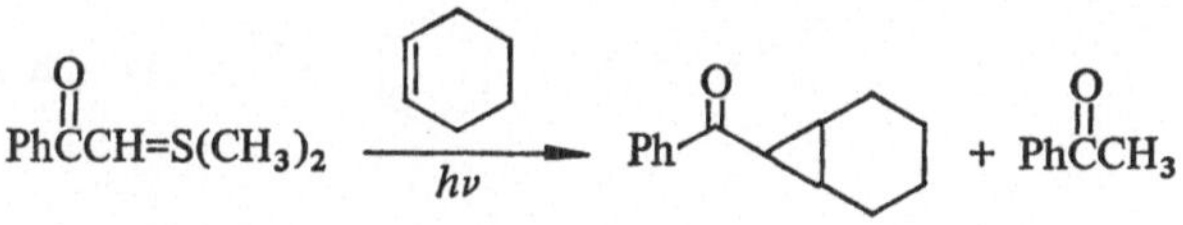

17

solvent) when irradiated in cyclohexene in the same ratio as the photo-decomposition of diazoacetophenone.[22] Most interesting, diphenylsulfo-niumallylide also suffers facile photodecomposition.[4] The major product was that expected for vinylcarbene, cyclopropene, which was isolated as its Diels-Alder adduct with spiro[4.2]hepta-2,4-diene in 25% yield. Synthe-

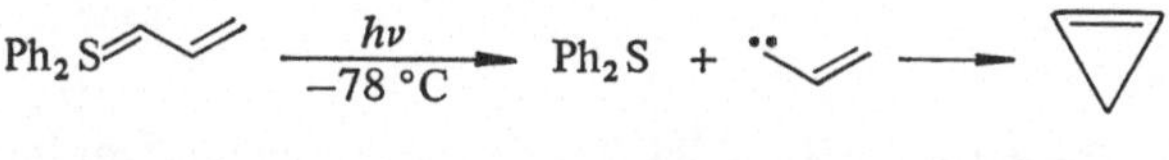

tically, this represents one of the best ways to generate the parent cyclo-propene by α elimination since vinyldiazomethane produces it in an infi-nitesimal yield and allylchloride in only 5—10% yield.[24]

Thus, sulfur ylides undergo α elimination quite well under photolytic conditions. Considering their ease of handling in comparison to the corre-sponding diazo compounds, they should find great application as carbene precursors.

11

IV. σ Sulfuranes

In the generation of diphenylsulfoniumethylide with organolithium, Corey reports the formation of biphenyl.[25] Generation of diphenylsulfoniumallylide with n-butyllithium is accompanied by biphenyl, allylbenzene, and n-butylphenyl sulfide.[4] In the latter case, these types of products were almost completely suppressed by use of the much more bulky t-butyllithium. The origin of these coupling products led to the consideration of σ sulfurane intermediates.

A. Arylsulfuranes

Wittig reported the isolation of biphenyl and diphenyl-sulfide upon treatment of triphenylsulfonium salt with phenyllithium.[26] Although the origin of these products may be *via* the decomposition of a sulfurane, a

$$\text{Ph}_3\overset{\oplus}{\text{S}} + \text{PhLi} \nearrow \text{Ph}_4\text{S} \longrightarrow \text{Ph--Ph} + \text{Ph}_2\text{S}$$

$$\searrow \text{Ph}_2\text{S} + \bigcirc \xrightarrow{\text{PhLi}} \text{Ph--Ph}$$

benzyne route cannot be ruled out. Such a process can be eliminated by consideration of the reaction of tri-p-tolylsulfonium fluoroborate with p-tolyllithium.[27] The observed coupling product retained the methyl label

$$\left(\text{CH}_3-\bigcirc-\right)_3 \overset{\oplus}{\text{S}} \overset{\ominus}{\text{BF}_4} + \text{CH}_3-\bigcirc-\text{Li} \longrightarrow \text{CH}_3-\bigcirc-\bigcirc-\text{CH}_3$$

$$+$$

$$\text{CH}_3-\bigcirc-\text{S}-\bigcirc-\text{CH}_3$$

exclusively in the *para* position.[28] If a benzyne mechanism had been operative, the major products should possess the methyl groups *meta* since 3,4-

dehydrotoluene is known to suffer nucleophilic addition predominantly in the *meta* position.[29]

An alternative possibility in accord with the above labeling results is a nucleophilic addition elimination. The reaction of phenyllithium and vinyllithium with S-phenyldibenzothiophenium fluoroborate addresses itself to this problem.[30] This salt reacts with phenyllithium to yield 2-phenylthio-*o*-terphenyl exclusively in nearly quantitative yields. If an addition elimina-

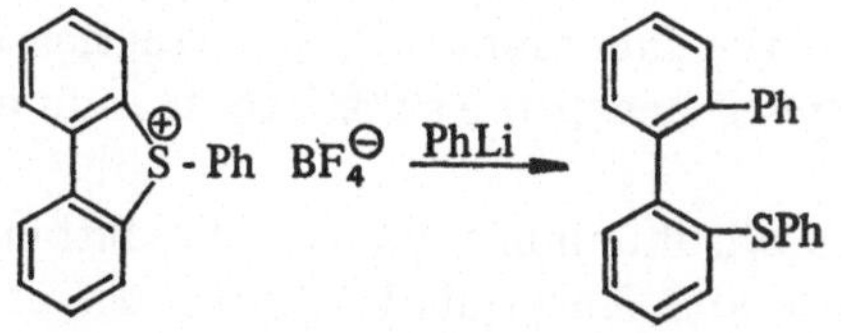

tion mechanism is operative, the adduct leading to this product must be *18a*. The complete preferential formation of *18* over *19* is quite under-

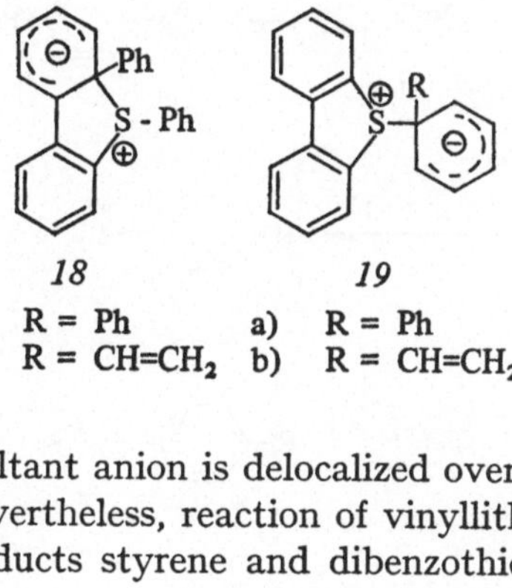

18

a) R = Ph
b) R = CH=CH$_2$

19

a) R = Ph
b) R = CH=CH$_2$

standable since the resultant anion is delocalized over two rings in *18*, but only one ring in *19*. Nevertheless, reaction of vinyllithium with the dibenzothiophenium salt products styrene and dibenzothiophene as the major

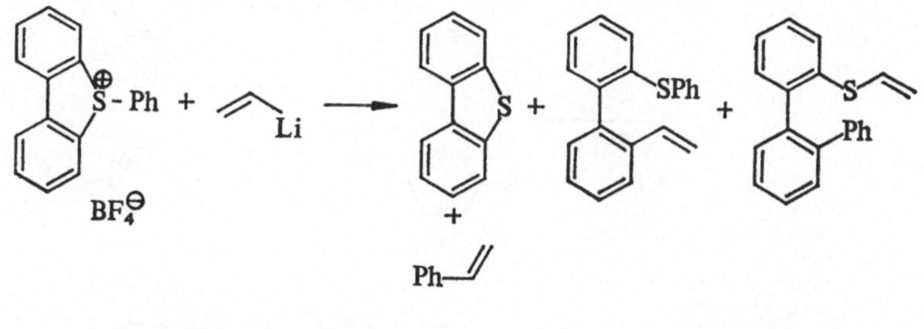

products (approximately 60% yield each) and 2-phenylthio-2'-vinylbiphenyl and 2-vinylthio-*o*-terphenyl as the minor products (approximately

13

6% yield each). This result requires *19b* to be highly preferred over *18b* if an addition-elimination mechanism is operative. Such a change in relative stability of *18* and *19* in changing from R=Ph to R=CH=CH$_2$ is not reasonable. The only mechanism that is consistent with all the results is a sulfurane route.

The utility of this process as a means of forming carbon-carbon bonds was briefly investigated.[27] Formation of biaryls upon treatment of triarylsulfonium salts occurs in high yields. Synthetically, it is most practical for symmetrical biaryls. The sulfide by-product can be recycled since it is the starting material for the sulfonium salt. The reaction can be applicable to unsymmetrical biaryls if one of the substituents is an electron withdrawing group.

Treatment of triarylsulfonium salts with vinyllithium leads to quantitative yields of styrene and diarylsulfide. Again, use of tri-*p*-tolylsulfonium fluoroborate produces only *p*-methylstyrene, ruling out benzyne routes.

Labeled vinyllithiums retain their stereochemical integrity. Thus, *cis*- and *trans*-propenyllithium lead to *cis*- and *trans*-propenylbenzene, respectively, with no crossover.

The reaction was extended to allyl-aryl coupling with excellent results. Treatment of the triarylsulfonium salt with allyllithium (generated from tetraallylstannane) produced allylarenes in quantitative yields. Nevertheless the reaction could not be extended to benzyllithium nor saturated alkyllithiums. In the former case, a complex reaction mixture which was not analyzed was obtained. The addition of *t*-butyllithium exemplifies the latter case. Only biaryl and aryl-*t*-butylsulfide was obtained. This result clearly indicates the requirement of a π system for coupling.

The mechanism for collapse of a sulfurane to these products remains obscure. Homolytic scission of two carbon sulfur bonds followed by coupling

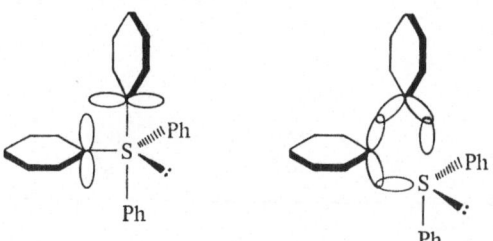

of the resultant free radicals can be ruled out. First, alkyl groups do not couple. Since alkyl radicals are more stable than aryl radicals, but the present results demand preferential formation of the latter, a clear contradiction with a radical mechanism exists. Second, in the propenyllithium reactions, the geometrical integrity of the vinyl group is retained. Since vinyl radicals are known to invert with a frequency of about 10^{10}/sec,[32] this result requires the coupling rate to be at least 10^{12}/sec. Fig. 1 pictures a concerted collapse and explains the requirement of a π system. Quantum mechanically a p plus a sp^2 orbital may be considered as two sp^5 orbitals. Thus, the reaction involves six orbitals and six electrons, a ground state allowed reaction.

Fig. 1. Rationale for coupling

To determine the nature of the electronic distribution in the two rings undergoing coupling, substituent effects were investigated.[31] S-Aryldibenzothiophenium salts were chosen as the substrates for study since they were readily available and since migratory aptitudes could be determined uncomplicated by ligand exchange [32] when the aryl groups of the S-aryl salt and the aryllithium are different. Two modes of decomposition of the sulfurane may be envisioned—aryl-biphenylyl coupling (path A) and aryl-aryl coupling (path B). The ratio of these two paths is a function of the nature of the aryl groups. Table 2 summarizes the ratio of path A to path B for the cases of $Ar = Ar'$. Whereas, aryl groups, unsubstituted or substituted

Table 2. Relative yields in symmetrical coupling reactions

Ar			
C_6H_5	—	—	100
$p-CH_3C_6H_4$	—	—	100
$m-CF_3C_6H_4$	20.0	26.6	53.4
$p-CF_3C_6H_4$	28.5	25.1	46.4

with electron donating groups react by aryl-biphenyl coupling exclusively, equal amounts of aryl-aryl and aryl-biphenylyl coupling occur with aryl groups substituted with electron withdrawing groups. Clearly, electron withdrawing groups facilitate the aryl coupling reaction.

16

This conclusion is heavily reinforced by consideration of the reaction when $Ar \neq Ar'$ (see Table 3). Not only is a similar trend seen in terms of the ratio of path A to path B, but also the internal competition in terms of the ratio of the two terphenyls demonstrates the facilitation of coupling by electron withdrawing substituents. Changing from the most electron donat-

Table 3. Relative yields in unsymmetrical coupling reactions

Ar¹	Ar²	Temp.				
p—CH₃C₆H₄	C₆H₅	− 78	—	—	94.5	5.5
C₆H₅	pCH₃C₆H₄	− 78	—	—	6.6	93.4
pCH₃O—C₆H₄	C₆H₅	−100	—	—	100	—
		− 78	—	—	100	—
		− 20	—	—	100	—
C₆H₅	pCH₃OC₆H₄	−100	—	—	—	100
		− 78	—	—	—	100
		− 20	—	—	—	100
pCF₃C₆H₄	C₆H₅	− 78	1.6	1.6	2.5	94.3
C₆H₅	pCF₃C₆H₄	− 78	2.4	2.4	92.1	3.2
mCF₃C₆H₄	C₆H₅	−100	18.2	18.2	—	63.6
		− 78	29.4	27.1	—	43.4
		− 20	30.5	24.6	—	44.9
C₆H₄	mCF₂C₆H₄	−100	9.2	8.7	82.1	—
		− 78	3.3	2.9	93.8	—
		− 20	3.9	5.0	91.1	—

ing substituent, *p*-methoxy, to the most electron withdrawing substituent, *m*-trifluoromethyl, a gradation from no migration to total migration of the substituted ring occurs. It may be noted that a biphenylyl group may be considered as an aryl group substituted in the *ortho* position by a phenyl substituent, an electron withdrawing substituent. Thus, coupling strongly prefers both aryl groups to bear electron withdrawing substituents.

17

The significance of such migratory preferences must be tempered by consideration of the geometry of the sulfurane *20*. By analogy to phosphorus chemistry [23], the following structural assumptions are made:

1) pentacoordinated sulfur exists as a trigonal bipyramid,
2) the electron pair prefers a basal orientation,
3) the five-membered ring prefers the apical-basal orientation, and
4) groups enter and leave preferentially from an apical position.

Examination of molecular models shows that aryl groups in the apical position cannot couple with the biphenyl unit without an inordinate amount of strain. Thus, sulfurane *20a*, which is the kinetic product of addition of Ar^1Li to S-Ar-dibenzothiophenium fluoroborate, gives rise to terphenyl *21*, and sulfurane *20b*, which is the kinetic product of addition of ArLi to

| 21 | 20a | 20b | 22 |

S-Ar'-dibenzothiophenium fluoroborate, gives rise to terphenyl *22*. That *20a* and *20b* interconvert faster than coupling and therefore the ratio of *21* and *22* reflects the relative migratory aptitudes of Ar and Ar' is indicated by two observations. First, the ratio of *21* to *22* is temperature independent over the range −100 °C to −20 °C. Second, the ratio does not depend on the order of addition of the aryl groups to form the sulfurane.

The terphenylaryl sulfides produced in this way are quantitatively desulfurized by Raney nickel in refluxing ethanol. In this way *o*-terphenyl

23

and many substituted *o*-terphenyls (*23*, $R = CH_3$, OCH_3, CF_3, $R' = H$ and $R = H$, $R' = CF_3$) free of other isomers were prepared.

Stereochemical effects on coupling can be seen if the rate of coupling is increased relative to sulfurane isomerization. The vinyl group undergoes coupling exceedingly efficiently. Thus, triphenylvinyl sulfurane generates styrene and no biphenyl.[29] Treatment of s-aryldibenzothiophenium fluoroborate with vinyllithium generates predominantly 2-arylthio-2'-vinylbiphenyls if the aryl group possesses electron donating substituents but styrenes and dibenzothiophene if the aryl group is unsubstituted or substituted with an electron with drawing group (see Table 4).[31]

Table 4. Relative yields of vinyl coupling

Ar				
Ph	41.0	41.0	11.0	7.0
pCH$_3$—C$_6$H$_4$	7.6	7.6	70.0	14.8
pCH$_3$OC$_6$H$_4$	3.5	3.8	80.5	12.2
mCF$_3$C$_6$H$_4$	50.3	48.1	—	1.6
pCF$_3$C$_6$H$_4$	48.8	48.3	—	2.9

The ratio of the two types of products are easily rationalized if the rate of coupling is competitive or somewhat faster than sulfurane interconversion. Thus, sulfurane *24* is the kinetic product. It can only undergo vinyl-aryl or aryl-biphenylyl coupling. The alternate geometrical isomer *25* can undergo vinyl-aryl or vinyl-biphenylyl coupling. The previous results indicate that the vinyl and biphenylyl groups participate in the coupling most efficiently. However, when the aryl group possesses electron withdrawing substituents or no substituents, the rapid rate of coupling precludes isomerization of *24* to *25*. The major products are thus dibenzothiophene and the substituted styrene. When the aryl group possesses electron donating substituents, vinyl-aryl coupling is slowed. Interconversion of *24* and *25* now occurs more rapidly than coupling and the expected vinyl-biphenylyl coupling predominates.

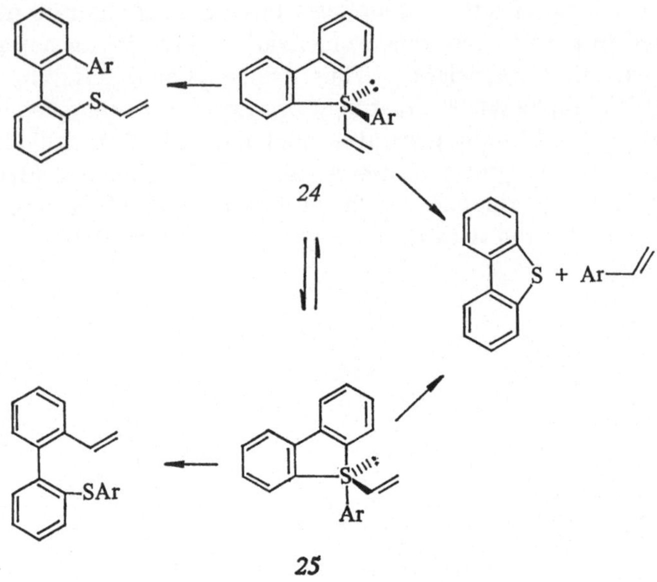

24

25

Further support for this interpretation arises by consideration of the reaction of propenyllithium with S-*m*-trifluoromethylphenyldibenzothiophenium fluoroborate. Just as in the aryl cases, alkyl substituents on the vinyl group hinder the coupling. Whereas, with sulfurane *24*, Ar = *m*-trifluoromethylphenyl, coupling occurred much faster than sulfurane interconversion giving dibenzothiophene and *m*-trifluoromethylstyrene in 98.4% yield, collapse of sulfurane *26* to dibenzothiophene and *m*-trifluoro-

26

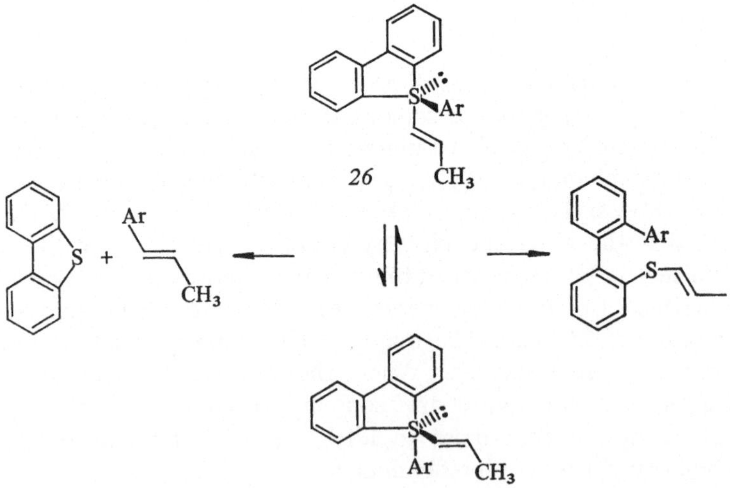

methylpropenylbenzene diminishes to 77.7%. The terphenyl, present in only 1.6% from *24* now *is* present in 22.3%.

Interpretation of the chemistry of this new arylation reaction in terms of sulfuranes received strong confirmation in the recent isolation of a stable sulfurane as a crystalline solid. Tetrakis-pentafluorophenylsulfurane, stable only below 0 °C, was formed by reaction of pentafluorophenyllithium and

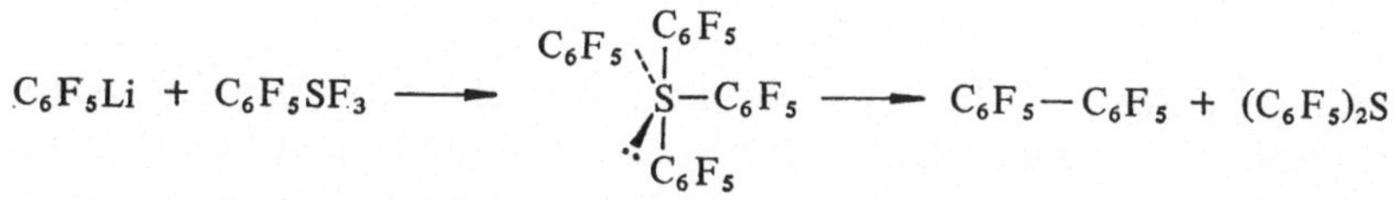

$$C_6F_5Li + C_6F_5SF_3 \longrightarrow \underset{\underset{\ddot{C_6F_5}}{|}}{\overset{\overset{C_6F_5}{|}}{\underset{C_6F_5}{S}}}{-}C_6F_5 \longrightarrow C_6F_5{-}C_6F_5 + (C_6F_5)_2S$$

pentafluorophenylsulfur trifluoride. Thermal decomposition produced perfluorobiphenyl and perfluorodiphenyl sulfide in equimolar quantities.

The utility of the arylsulfonium salts as arylating agents in the above reactions depends upon the availability of these salts. Triphenylsulfonium bromide is commercially available. Condensation of organometallics with diarylsulfoxides [34] or diarylethoxysulfonium fluoroborates [35] suffers from low yields. Use of diaryladamantoxysulfonium fluoroborate circumvents

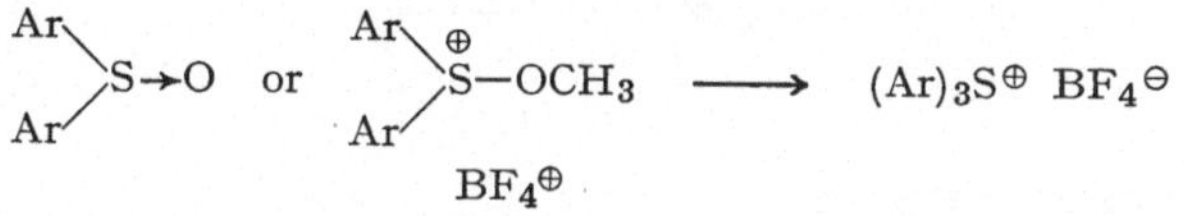

$$\underset{Ar}{\overset{Ar}{>}}S{\rightarrow}O \quad or \quad \underset{\underset{BF_4^{\oplus}}{Ar}}{\overset{Ar}{>}}\overset{\oplus}{S}{-}OCH_3 \longrightarrow (Ar)_3S^{\oplus}\ BF_4^{\ominus}$$

this problem.[36)] In this way, triphenylsulfonium fluoroborate was isolated in over 90% yield. The adamantanol by-product may be recycled.

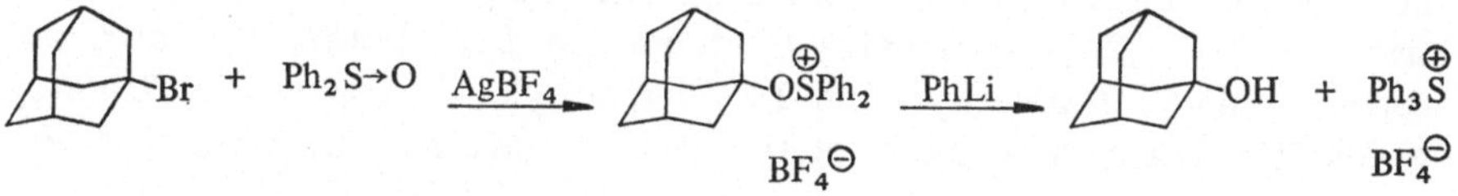

$$\text{(adamantyl)}{-}Br + Ph_2S{\rightarrow}O \xrightarrow{\ AgBF_4\ } \text{(adamantyl)}{-}\overset{\oplus}{O}SPh_2 \xrightarrow{\ PhLi\ } \text{(adamantyl)}{-}OH + Ph_3\overset{\oplus}{S}$$
$$BF_4^{\ominus} \qquad\qquad BF_4^{\ominus}$$

B. Alicyclic Sulfuranes

1. Three-Membered Rings

Bordwell and coworkers reported the stereospecific quantitative desulfurization of episulfides with organolithiums.[39)] The mechanism of this reaction, however, remained undefined. In addition to a disrotatory concerted sulfurane decomposition (*i.e.* Eq. 9), β elimination from a 2-alkylthioalkyl-lithium may be envisioned (see Eq. 10). However, independent stereospecific generation of this organolithium from the corresponding 1-bromo-2-alkylthioethers demonstrated the nonstereospecificity of the elimination reaction.[38] Thus, episulfide desulfurization represents the earliest example of an aliphatic sulfurane.

2. Four-Membered Rings

The desulfurization of a thietane would create a new cyclopropane synthesis. Although episulfides react quantitatively with organolithiums at low temperatures, thietanes are unreactive. However, alkylation of thietanes with Meerwein's reagent generates the corresponding thietanonium fluoroborates which readily react with organolithiums.[39] Treatment of 1,2,4-trimethyl- or 1,2,2,4-tetramethylthietanonium fluoroborate produces the corresponding 1,2-dimethyl- and 1,1,2-trimethylcyclopropanes in 20—30% yields. The reaction is highly stereospecific. Thus, cis-1,2,4-trimethylthie-

tanonium fluoroborate produces cis- and trans-1,2-dimethylcyclopropane in the ratio of 1:11 at −78 °C; whereas, trans-1,2,4-trimethylthietanonium fluoroborate produces the same cyclopropanes in a ratio 8:1 at −78 °C. The results are in best accord with sulfurane intermediates.

cis	1	11
trans	8	1

Addition of the butyl group to sulfur generates the sulfurane and initiates the decomposition process. Least motion collapse expels *n*-butyl-

methyl sulfide and produces the 1,3-diradical. Extended Hückel calculations [41] suggest and studies of pyrazoline decompositions [42] confirm that such trimethylenes undergo preferential conrotatory closure to the cyclopropanes. The extent of stereospecificity observed here far exceed that observed in thermal pyrazoline decomposition — an observation attributable to the nearly 200 °C temperature difference of the two reactions.

3. Five-Membered Rings

The sulfurane from a 2,5-dihydrothiophenium salt should undergo preferential fragmentation to a 1,3-diene and a dialkylsulfide. Furthermore, orbital symmetry dictates a disrotatory fragmentation. In principle, such inter-

mediates are available by the addition of an alkyllithium to a dihydrothiophenium salt. 2,5-Dihydrothiophene previously has been available by the dissolving metal reduction of thiophene; however, the yields were very low and the reaction was not applicable to substituted thiophenes. [43] These compounds were available by the two step sequence outlined below. [44] Concurrent dropwise addition of hexane solutions of a diene and sulfur dichloride to a reservoir of hexane produced the corresponding 3,4-dichloro-tetrahydrothiophenes as a mixture of isomers. [45] Reductive elimination succeeded quantitatively at room temperature in aqueous dimethylformamide with chromous ion [46] to the corresponding 2,5-dihydrothiophenes.

R	R^1	R^2	R^3
CH$_3$	H	CH$_3$	H
CH$_3$	CH$_3$	CH$_3$	CH$_3$
CH$_3$	C$_2$H$_5$	CH$_3$	C$_2$H$_5$

It is interesting to note that *trans,trans*-2,4-hexadiene gives 2,5-dimethyl-2,5-dihydrothiophene as a 2:3 mixture of the *cis* and *trans* isomers, whereas, a ratio of 1:9 is obtained from *cis,trans*-2,4-hexadiene. The stereospecificity of the addition of sulfur dichloride may be attributable to a concerted 1,4-addition followed by chlorine migration to the double bond. The high

electrophilicity of sulfur dichloride and the favorability of a five-membered ring in a trigonal bipyramide geometry make such a pathway attractive. An analogous reaction has been observed in phosphorus chemistry.[49]

Dihydrothiophenes themselves are inert to *n*-butyllithium. They were readily convertible into their corresponding sulfonium salts by methylation with trimethyloxonium fluoroborate in acetonitrile or methylene chloride. Metathesis with ammonium hexafluorophosphate converted the hygroscopic fluoroborate salts into the nonhygroscopic hexafluorophosphates.

Treatment of the 2,2,5,5-tetraalkyldihydrothiophenium salts with *n*-butyllithium produced the corresponding dienes and *n*-butylmethyl sulfide in only low yields. The major products arose by E$_2$ elimination. On the other hand, decreasing the steric hindrance to attack at sulfur greatly facilitated the fragmentation reaction. Thus, the 2,5-dimethyl-2,5- dihydrothiophenium salts led to fragmentation products in yields of 70—90%. Furthermore, the fragmentation reaction is completely stereospecific clearly implicating the sulfurane intermediate (see Scheme 1). Thus, *cis*-1,2,5-trimethyl-2,5-dihydrothiophenium hexafluorophosphate gave only *cis,cis*-

and *trans,trans*-2,4-hexadienes in which the former predominated. Employing *t*-butyllithium or phenyllithium only the *trans,trans* isomer was obtained. With *trans*-1,2,5-trimethyl-2,5-dihydrothiophenium hexafluorophosphate

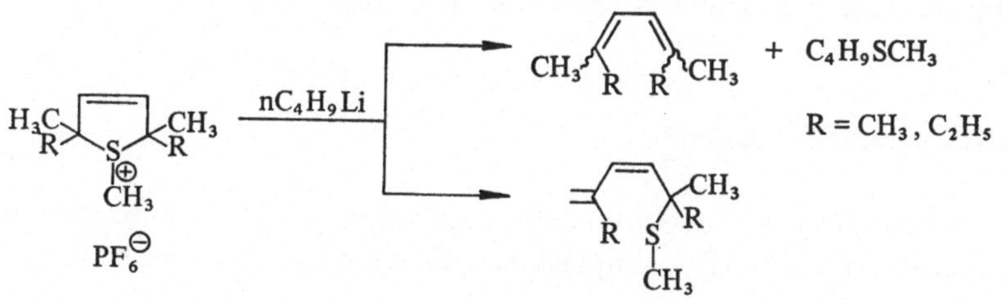

only *cis,trans*-2,4-hexadiene was obtained. Thus, fragmentation involves clean disrotatory motion. A minor product in both reactions was 2-methylthio-2,4-hexadiene. It arises by elimination in the allylide derived from the sulfonium salts. This compound is the sole product when non-nucleophilic bases like potassium *t*-butoxide are employed.

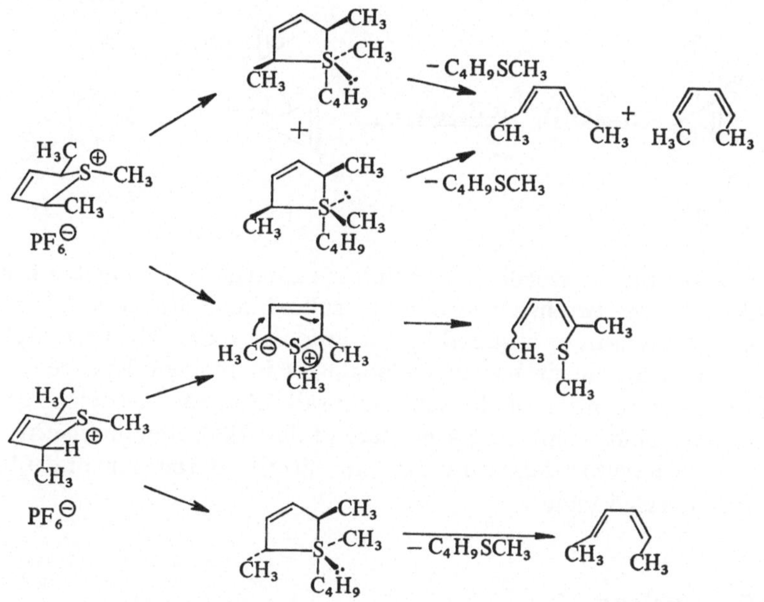

Scheme 2. Fragmentation of 2,5-dihydrothiophenium hexafluorophosphates

One of the most fascinating features of this reaction is the high preference for σ sulfurane formation over π sulfurane formation even though the latter is a stabilized π sulfurane. Westheimer has pointed out that placing phosphorus in a five-membered ring enhances phosphorane formation compared to acyclic or larger ring sizes.[33] The observation has been attributed to the fact that the bond angles of the five-membered ring are better accommodated when phosphorus is trigonal bipyramide compared to it being tetrahedral. It appears a similar effect is operative in sulfur chemistry.

V. Thiophene Synthesis

The reaction of cyclopropenium salts with sulfur ylides leads to thiophenes in moderate yields.[48] The simplest rationale for this most interesting reac-

tion invokes the intermediacy of sulfurane 28. The stoichiometry has been established to be two moles of dimethylsulfoniummethylide to one mole of cyclopropenium salt as required by the proposed route. The thiavinylcyclopropene rearrangement has many analogies in iminocyclopropene chemistry.[49] Deprotonation of the sulfonium salt 27 at the methylene group is irreversible. Thus, employing a mixture of dimethylsulfoniummethylide-d$_8$ gave thiophene with a deuterium content reflective of the ratio of deuterated to nondeuterated ylide.

VI. Conclusions

The involvement of sulfuranes in the reactions of organosulfur compounds is only now coming to be realized. Classic reactions such as sulfide oxidations[50] and sulfenyl chloride additions to unsaturated linkages [51] are claimed to involve such intermediates. The products from the reactions of heteroatom nucleophiles with triarylsulfonium salts appear to have arisen from σ sulfurane intermediates.[52] A diphenyldialkoxysulfurane serves as a new mild dehydrating agent.[53,54] Most fascinating is the role σ and π sulfuranes may play as a method for generating new carbon-carbon bonds. These investigations begin to demonstrate that such intermediates are versatile new synthetic reagents and are involved in many reactions of organosulfur compounds. Such intermediates will play an increasingly important role in the development of new synthetic methods.

Acknowledgment. I want to thank my very able coworkers who carried out this work — Dr. Robert C. Atkins, Dr. Ronald W. LaRochelle, Mr. Mitchell Bogdanowicz, and Mr. Henry C. Arndt. I want to thank the National Institutes of Health and the National Science Foundation for their generous support.

VII. References

[1] For a recent summary of some π sulfurane reactions, see House, H. O.: Modern synthetic reactions, 2nd edit. Menlo Park, Calif.: W. A. Benjamin Inc. 1972. — Field, L.: Synthesis, *3*, 101 (1972). — Agami, C.: Bull. Soc. Chim. France *1965* 1021. — Johnson, A. W.: Ylid chemistry. New York: Academic Press 1966.

[2] Sulfonium Ylides:
   a) Corey, E. J., Chaykovsky, M.: J. Am. Chem. Soc. *87*, 1353 (1956);
   b) LaRochelle, R. W., Trost, B. M., Krepski, L.: J. Org. Chem. *36*, 1126 1971);
   c) Franzen, V., Driesen, H. E.: Chem. Ber. *96*, 1881 (1963).

[3] Oxosulfonium Ylides:
   a) Corey, E. J., Chaykovsky, M.: J. Am. Chem. Soc. *87*, 1353 (1965);
   b) Johnson, C. R., Haake, M., Schroeck, C. W.: J. Am. Chem. Soc. *92*, 6594 (1970);
   c) Johnson, C. R., Kateker, G. F.: J. Am. Chem. Soc. *92*, 5753 (1970).

[4] LaRochelle, R. W., Trost, B. M., Krepski, L.: J. Org. Chem. *36*, 1126 (1971).

[5] Trost, B. M., Bogdanowicz, M. J.: J. Am. Chem. Soc. *93*, 3773 (1971).

[6] Bogdanowicz, M. J., Trost, B. M.: Tetrahedron Letters *1972*, 887.

[7] Johnson, C. R., Katekar, G. F., Huxol, R. F., Janiga, E. R.: J. Am. Chem. Soc. *93*, 3771 (1971).

[8] Horton, J. A., Laura, M. A., Kalberg, S. M., Petterson, R. C.: J. Org. Chem. *34*, 3366 (1969).

[9] Trost, B. M., Ambelang, T., Bogdanowicz, M. J.: Tetrahedron Letters, 923 (1973).

[10] Tsuda, Y., Tanno, T., Vkai, A., Isobe, K.: Tetrahedron Letters, *22*, 2009 (1971) .— Corey, E. J., Ravindranathan, T.: Tetrahedron Letters *49*, 4753 (1971).

[11] Trost, B. M., Bogdanowicz, M. J.: J. Am. Chem. Soc., *95*, 2038 (1973).

[12] Walborsky, H. M., Allen, L. E.: J. Am. Chem. Soc. *93*, 5465 (1971).

[13] Trost, B. M., Bogdanowicz, M. J.: J. Am. Chem. Soc. *94*, 4777 (1972).

[14] For a review of fragmentation reactions, see Grob, C. A., Schiess, P. W.: Angew. Chem. Intern. Ed. Engl. *6*, 1 (1967). — Grob, C. A.: Angew. Chem. Intern. Ed. Engl. *8*, 535 (1969). — Marshall, J. A.: Rec. Chem. Prog. *30*, 3 (1969); Synthesis *1971*, 229.

15) Wharton, P. S., Hiegel, G. A.: J. Org. Chem. *30*, 3254 (1965). — Wharton, P. S., Hiegel, G. A., Coombs, R. V.: J. Org. Chem. *29*, 3217 (1963). — Wharton, P. S.: J. Org. Chem. *26*, 4781 (1961).

16) Bergmann, E. D., Ginsburg, D., Pappo, R.: Org. Reactions *10*, 179 (1969).

17) Franzen, V., Schmidt, H. J., Mertz, C.: Chem. Ber. *94*, 2942 (1962).

18) Johnson, A. W., Hruby, V. J., Williams, J. L.: J. Am. Chem. Soc. *85*, 918 (1964).

19) Swain, C. G., Thornton, E. R.: J. Org. Chem. *26*, 4808 (1961).

20) Corey, E. J., Chaykovsky, M.: J. Am. Chem. Soc. *86*, 1640 (1964).

21) Kunieda, T., Witkop, B.: J. Am. Chem. Soc. *93*, 3487 (1971).

22) Trost, B. M.: J. Am. Chem. Soc. *89*, 138 (1967). Also see Johnson, A. W., Amel, R. T.: J. Org. Chem. *34*, 1240 (1969).

23) Closs, G. L.: In: Advances in alicyclic chemistry (ed. H. Hart and G. J. Karabatsos), Vol. I, pp. 63—64. New York: Academic Press 1966.

24) Closs, G. L., Krantz, K. D.: J. Org. Chem. *31*, 638 (1966). See also Magid, R. M., Welch, J. G.: J. Am. Chem. Soc. *90*, 5211 (1968).

25) Corey, E. J., Jautelat, M.: J. Am. Chem. Soc. *89*, 3912 (1967).

26) Wittig, G., Fritz, H.: Liebigs Ann. Chem. *577*, 39 (1952). For a modified reaction, see Sheppard, W.: J. Am. Chem. Soc. *84*, 3058 (1962).

27) LaRochelle, R. W., Trost, B. M.: J. Am. Chem. Soc. *93*, 6077 (1971). — Trost, B. M., LaRochelle, R. W., Atkins, R. C.: I. Am. Chem. Soc. *91*, 2175 (1969).

28) For related studies, see Khim, Y. H., Oae, S.: Bull. Soc. Chem. Japan *42*, 1968 (1969). — Andersen, K. K., Yeager, S. A., Peynircioglu, N. B.: Tetrahedron Letters *1970*, 2485.

29) Friedman, L., Chlebowski, J. F.: J. Am. Chem. Soc. *91*, 4864 (1969). — Scardiglia, F., Roberts, J. D.: Tetrahedron *3*, 197 (1968). — Ridegraff, G. B., den Hertog, J. J., Melger, W. C.: Tetrahedron Letters *1965*, 963.

30) Kampmeier, J. A., Fantazier, R. M.: J. Am. Chem. Soc. *88*, 1959 (1966).

31) Trost, B. M., Arndt, H. C.: J. Am. Chem. Soc., in press.

32) Franzen, V., Mertz, C.: Liebigs Ann. Chem. *643*, 24 (1961); Angew. Chem. *72*, 416 (1960).

33) For a discussion of phosphoranes, see Gorenstein, D., Westheimer, F. H.: J. Am. Chem. Soc. *92*, 634 (1970). — Westheimer, F.: Accounts Chem. Res. *1*, 70 (1968). — Gorenstein, D.: J. Am. Chem. Soc. *92*, 644 (1970).

34) Wildi, B. S., Taylor, S. W., Potratz, H. A.: J. Am. Chem. Soc. *73*, 1965 (1951).

35) Andersen, K. K., Papanikolaow, N. E.: Tetrahedron Letters *1966*, 5445. — Andersen, K. K., Cinquini, M., Papanikolaow, N. E.: J. Org. Chem. *35*, 706 (1970).

36) Trost, B. M., Hammen, R.: Unpublished results.

37) Neureiter, N. P., Bordwell, F. G.: J. Am. Chem. Soc. *81*, 578 (1959). — Schuetz, R. D., Jacobs, R. L.: J. Org. Chem. *26*, 3467 (1961).

38) Trost, B. M., Ziman, S. D.: Chem. Commun. *1969*, 181: J. Org. Chem. *38*, 649 (1973).

39) Trost, B. M., Schinski, W. L., Chen, F., Mantz, I. B.: J. Am. Chem. Soc. *93*, 676 (1971).

40) For an earlier claim of thietane decomposition with *n*-butyllithium, see Bordwell, F. G., Andersen, H. M., Pitt, B. M.: J. Am. Chem. Soc. *76*, 1982 (1954).

41) Hoffmann, R.: J. Am. Chem. Soc. *90*, 1475 (1968).

42) Crawford, R. J., Mishra, A.: J. Am. Chem. Soc. *88*, 3963 (1966). — Moore, R., Mishra, A., Crawford, R. J.: Can. J. Chem. *46*, 3305 (1968).

43) Birch, S. F., McAllan, D. T.: Nature *165*, 899 (1950). — King, R. C., Tocker, S.: J. Org. Chem. *20*, 1 (1955).

44) Trost, B. M., Ziman, S. D.: J. Am. Chem. Soc. *93*, 3825 (1971).

45) Compare Baker, H. J., Strating, J.: Rec. Trav. Chim. *154*, 52 (1935). — Lautenschlaeger, F.: J. Org. Chem. *31*, 1669 (1966).

[46] Kochi, J. K., Singleton, D. M., Andrews, L. J.: Tetrahedron *24*, 3503 (1968).

[47] Bond, A., Green, M., Pearson, S. C.: J. Chem. Soc. B *1968*, 929.

[48] Trost, B. M., Atikins, R.: Tetrahedron Letters *1968*, 1225; J. Am. Chem. Soc. *95*, 1285 (1973).

[49] See for example, Breslow, R., Boikess, R., Battiste, M.: Tetrahedron Letters *1960*, 42.

[50] Johnson, C. R., Rigau, J. J.: J. Am. Chem. Soc. *91*, 5398 (1969).

[51] Calo, V., Scorrano, G., Modena, G.: J. Org. Chem. *34*, 2020 (1969).

[52] Knapczyk, J. W., McEwen, W. E.: J. Am. Chem. Soc. *91*, 145 (1969). — Wiegand, G. H., McEwen, W. E.: J. Org. Chem. *33*, 2671 (1968).

[53] Martin, J. C., Arhart, R. J.: J. Am. Chem. Soc. *93*, 2339, 2341, 4327 (1971).

[54] Arhart, R. J., Martin, J. C.: J. Am. Chem. Soc. *94*, 4997, 5003 (1972).

Received August 25, 1972

Electron Correlation and Electron Pair Theories*

Prof. Dr. Werner Kutzelnigg

Institut für Physikalische Chemie und Elektrochemie der Universität Karlsruhe, Abteilung Theoret. Chemie, Karlsruhe**

Contents

---

* Slightly expanded version of a lecture given at the summer school on Computational Quantum Chemistry, Sept. 1971 in Ramsau/Berchtesgaden, Germany. The author is grateful to the organizers, particularly to Dr. G. H. F. Diercksen for the opportunity to give this lecture.

** Present address: Lehrstuhl für Theoret. Chemie der Ruhruniversität, Bochum.

31

I. Introduction

A. History and Definitions

The term "electron correlation" was coined by Wigner and Seitz [1] in the context of their study of the electronic structure and cohesive energy of metals. Two different aspects of correlation that are now familiar can be traced back to this first reference. On the one hand, "correlation energy" is defined as that part of the energy that one ignores when using a single Slater-determinant wave function. On the other hand, there is a statistical correlation of electrons in space as a consequence of the antisymmetry of the wave function. This "Fermi correlation", which prevents electrons with the same spin from coming too close to each other, has, however, very little to do with the "correlation energy", since the effect of Fermi correlation on the energy is allowed for in the energy of a Slater-determinant wave function. For an atom the difference between the Hartree and the Hartree-Fock energies might be referred to as Fermi correlation energy; the term "exchange energy" is, however, more usual.

The persisting semantic dilemma concerning "electron correlation" is that, with respect to correlation of electrons in space, Fermi correlation is regarded as part of the overall correlation, whereas with respect to correlation energy it is not.

The definition of correlation energy that refers to a single Slater-determinant wave function depends on the choice of the latter. If we seek a definition independent of such a choice, it is appropriate to refer to the "best" single Slater-determinant in terms of the energy criterion, i.e. the Hartree-Fock energy. It is further not correct to regard the whole error of the Hartree-Fock energy relative to the experimental energy as being due to correlation. It is necessary to distinguish relativistic effects and the like. A better reference than the experimental energy is thus the "exact non-relativistic energy", i.e. the appropriate eigenvalue of the nonrelativistic Schrödinger equation. Following Löwdin [2], it is now common to define the correlation energy as the difference between the "true" Hartree-Fock and the "true" non-relativistic energy. This definition, though apparently straightforward, has the disadvantage of being based on two quantities, neither of which can by definition be known exactly. Furthermore, for open-shell states there is no unique definition of the Hartree-Fock energy so, that different definitions of correlation energy are possible. Here, however, we shall consider mainly closed-shell states.

B. Accuracy of Correlation Energies

The correlation energy can only be estimated. One method is to start from some "near-Hartree-Fock" calculation and the corresponding experimental energy, extrapolate the former to the "Hartree-Fock limit" and correct the latter for relativistic effects in order to obtain an estimate of the exact non-relativistic energy. This yields an estimate of the "true" (or "experimental") correlation energy of a given state, which may then be compared with a "computed" correlation energy. Comparisons of this kind are never free from criticism and we may mention a recent controversy about a particular "computed correlation energy" for the neon ground state and its comparison with its experimental counterpart where an error of sign in the "Lamb shift" (*i.e.* a quantum electrodynamical correction) played a crucial part [3]. Our knowledge of relativistic (and related) effects for many-electron systems is too poor to allow us to make reliable estimates of "true non-relativistic energies", except for the very light atoms (say, those of the first period). Moreover, the extrapolation to the Hartree-Fock limit is somewhat uncertain, so that errors of a few per cent in "experimental" correlation energies are only to be expected. Nonetheless, the errors of computed correlation energies are at present somewhat larger, at least for molecules, so that "experimental" correlation energies can still serve as a reference standard for computed correlation energies. It is particularly useful to compare "experimental" and "computed" correlation energies for known molecules in order to check the reliability of the correlation energies predicted for unknown molecules.

C. Importance of Correlation Energy for Quantum Chemical Predictions

The correlation energy of an atom or molecule is usually of the order of 1% of the total energy, this being the energy required for complete atomization and ionization. (The zero of the energy scale corresponds to all nuclei and electrons at infinite distance from each other.) As long as we are interested in total energies, the correlation error is relatively small. But we are also interested in energy differences, which are small compared to the correlation energy. This is illustrated by Fig. 1. The relevant energy difference may be a spectral transition energy, a binding energy, a rotational barrier, etc.

With the example illustrated in Fig. 1, a "near-Hartree-Fock" calculation or an extrapolation of the Hartree-Fock limit does not give the correct sign for (E_1-E_2), since the change in correlation energy is larger than the quantity E_1-E_2 to be calculated. This is no exaggeration, examples of this kind are known. For example, a Hartree-Fock calculation of the binding of two F atoms to give the F_2 molecule yields the wrong sign for the binding energy. Of course, the correlation error is not the only error in any Hartree-Fock-type calculation. The relativistic corrections and the errors due to the limitation of the basis are rather large, too. We have good reason to believe that relativistic effects exert their main influence on the inner-shell electrons and that they do not change appreciably on molecule

Method	E_1, Molecule 1	E_2, Molecule 2

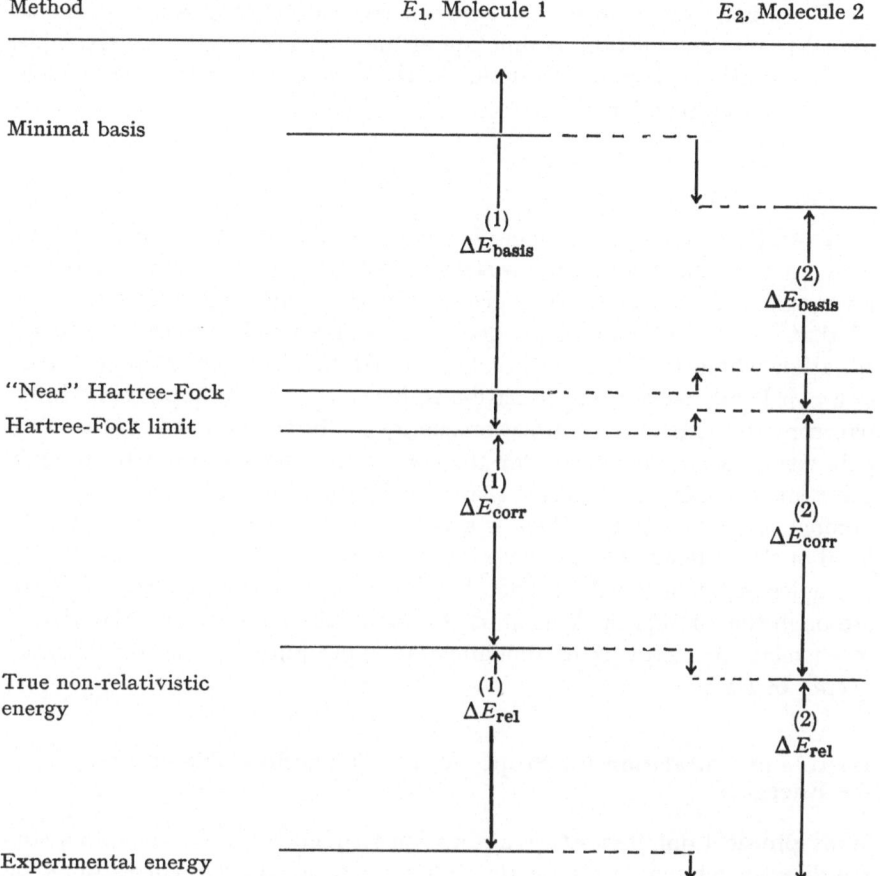

Fig. 1. Energies and energy differences calculated at different levels of approximation

formation or upon excitation of the outer shells. This has not been definitely proven but is commonly accepted. It is not yet possible to assess this question in a quantitative way.

Basis effects, on the other hand, are known to play an important role. In order to illustrate their importance we show in Fig. 1 a large change in the "basis" error on going from E_1 to E_2. The only way to avoid this kind of situation is to choose large and flexible basis sets in order to come sufficiently close to the Hartree-Fock limit. Otherwise there is a tendency to attribute to electron correlation energy effects which have nothing to do with it.

There is a class of reactions whose correlation energy is commonly assumed to remain constant as function of the nuclear coordinates. Examples are the rotational barrier of ethane [4], the interaction of two water molecules [5], or the reaction

$$NH_3 + HCl \longrightarrow NH_4^+ + Cl^- \quad [6]$$

In all these cases the number of coupled electron pairs remains constant. Now, in the standard picture of electron correlation the correlation energy is the sum of the correlation energies of individual pairs, it is therefore assumed that the correlation energy does not change much, in contrast to the situation where the reaction involves unpaired electrons forming a pair, or a pair being broken. The situation is, however, not so simple, since interpair contributions to the correlation energy can be as important as the intrapair terms. So the condition for the correlation energy to remain roughly constant is somewhat stronger than merely the requirement of a constant number of coupled pairs. There is also the requirement that the neighbourhood of the different pairs remain roughly the same. This is the case for the examples just mentioned. It is not, however, for the reaction $2\,BH_3 \rightarrow B_2H_6$, although the number of electron pairs remains the same. Indeed the, change in correlation energy is of fundamental importance for the dimerisation energy of BH_3 [7].

D. Role of Correlation for Properties not Depending Directly on the Energies

Many physical quantities (*e.g.* absorption frequencies, force constants etc.) are directly related to the energy and depend on electron correlations via the correlation energy.

There are other quantities that are related to expectation values other than that of the Hamiltonian (*e.g.* dipole and higher multipole moments, spin densities, field gradients). Most of the operators that come into play here are one-electron operators, and the Møller-Plesset theorem [8] states that their expectation values (like the electron density) are affected by correlation corrections to the wave function only to second order. As a consequence, correlation does not much influence these expectation values, except when the Hartree-Fock contribution is unusually small, so that a correlation correction may even determine the sign, as is the case for the dipole moment of CO [9].

A third class of quantities are calculated by second-order perturbation theory via a one-particle perturbation operator (*e.g.* electric polarizabilities, Van-der-Waals constants, diamagnetic susceptibilities, chemical shifts). These are affected by correlation effects to first order and hence depend sensitively on correlation, though this aspect has not been much investigated so far [9a].

II. Electron Correlation in the Statistical Sense

A. Density and Pair Density

"Correlation" is a technical term borrowed from probability theory, and it is appropriate to analyze electron correlation in this framework. All the information about electron correlation is contained in two functions that can be derived from the wave functions, namely electron density $\varrho(\vec{r})$ and pair density $\pi(\vec{r}_1, \vec{r}_2)$. We use this notation proposed by Ruedenberg [10] instead of the original notation $P_1(\vec{r})$ and $P_2(\vec{r}_1, \vec{r}_2)$ of McWeeny [11], as it cuts out some of the subscripts.

$$\varrho(\vec{r}_1) = n \int |\Psi(\vec{r}_1, \vec{r}_2 \ldots \vec{r}_n)|^2 \, d\tau_2 \ldots d\tau_n \cdot ds_1 \, ds_2 \ldots ds_n \tag{1}$$

$$\pi(\vec{r}_1, \vec{r}_2) = n(n-1) \int |\Psi(\vec{r}_1, \vec{r}_2, \ldots \vec{r}_n)|^2 \, d\tau_3 \ldots d\tau_n \, ds_1 \, ds_2 \ldots ds_n \tag{2}$$

Any $d\tau_i$ stands for the volume element in space for the i-th electron, ds_i refers to the integration over the spin coordinate. Both ϱ and π are spinless quantities. Expectation values of any multiplicative one- or two-electron operator can be expressed in terms of ϱ or π. If $e.g.$

$$A = \sum_{k=1}^{n} f(\vec{r}_k) \tag{3}$$

$$B = \sum_{k<l=1}^{n} g(\vec{r}_k, \vec{r}_l) \tag{4}$$

then

$$<\Psi|A|\Psi> = \int \varrho(\vec{r}) \, f(\vec{r}) \, d\tau_1 \tag{5}$$

$$<\Psi|B|\Psi> = \tfrac{1}{2} \int \pi(\vec{r}_1, \vec{r}_2) \, g(\vec{r}_1, \vec{r}_2) \, d\tau_1 \, d\tau_2 \tag{6}$$

For some applications, $e.g.$ the calculation of expectation values of non-multiplicative operators, like kinetic energy, one needs the density matrices $\varrho(\vec{r}_1, \vec{r}_1')$ and $\pi(\vec{r}_1, \vec{r}_2, \vec{r}_1', \vec{r}_2')$ rather than the density functions.

$$\varrho(\vec{r}_1, \vec{r}_1') = n \int \Psi(\vec{r}_1, \vec{r}_2 \ldots \vec{r}_n) \, \Psi^*(\vec{r}_1', r_2 \ldots \vec{r}_n) \, d\tau_2 \ldots d\tau_n \, ds_1 \ldots ds_n \tag{7}$$

$$\pi(\vec{r}_1, \vec{r}_2, \vec{r}_1', \vec{r}_2') = n(n-1) \int \Psi(\vec{r}_1, \vec{r}_2, \vec{r}_3 \ldots \vec{r}_n) \, \Psi^*(\vec{r}_1', \vec{r}_2', \vec{r}_3 \ldots \vec{r}_n) \, d\tau_3 \ldots d\tau_n$$
$$ds_1 \ldots ds_n \tag{8}$$

W. Kutzelnigg

The density functions are just the "diagonal elements" of the corresponding density matrices. Note that we use the same symbol for a density function and the corresponding density matrix and that

$$\varrho(\vec{r}) \equiv \varrho(\vec{r}, \vec{r}) \tag{9}$$

$$\pi(\vec{r}_1, \vec{r}_2) \equiv \pi(\vec{r}_1, \vec{r}_2; \vec{r}_1, \vec{r}_2) \tag{10}$$

We make use of the density matrix π to point out a general theorem which is useful with respect to the corresponding density function. We can decompose $\pi(\vec{r}_1, \vec{r}_2; \vec{r}_1', \vec{r}_2')$ into two components [12], [13] π_s and π_t

$$\pi = \pi_s + \pi_t \tag{11}$$

such that π_s is symmetric with respect to exchange of either $\vec{r}_1$ with $\vec{r}_2$ or $\vec{r}_1'$ with $\vec{r}_2'$ and π_t is antisymmetric with respect to the same exchange. Both π_s and π_t as well as π itself are symmetric with respect to *simultaneous* exchange of $\vec{r}_1$ with $\vec{r}_2$ and $\vec{r}_1'$ with $\vec{r}_2'$.

$$
\begin{aligned}
\pi_s(\vec{r}_1, \vec{r}_2; \vec{r}_1', \vec{r}_2') &= \pi_s(\vec{r}_2, \vec{r}_1; \vec{r}_1', \vec{r}_2') \\
= \pi_s(\vec{r}_1, \vec{r}_2; \vec{r}_2', \vec{r}_1') &= \pi_s(\vec{r}_2, \vec{r}_1; \vec{r}_2', \vec{r}_1') \\
\pi_t(\vec{r}_1, \vec{r}_2; \vec{r}_1', \vec{r}_2') &= -\pi_s(\vec{r}_2, \vec{r}_1; \vec{r}_1', \vec{r}_2') \\
= -\pi_t(\vec{r}_1, \vec{r}_2; \vec{r}_2', \vec{r}_1') &= \pi_t(\vec{r}_2, \vec{r}_1; \vec{r}_2', \vec{r}_1') \\
\pi(\vec{r}_1, \vec{r}_2; \vec{r}_1', \vec{r}_2') &= \pi(\vec{r}_2, \vec{r}_1; \vec{r}_2', \vec{r}_1') \\
\pi(\vec{r}_1, \vec{r}_2) &= \pi(\vec{r}_2, \vec{r}_1)
\end{aligned}
\tag{12}
$$

One can show [11,12] that the respective traces of π_s and π_t are related to the number n of particles and the quantum number S of the total spin through

$$\text{Tr } \pi_s = \int \pi_s(\vec{r}_1, \vec{r}_2) \, d\tau_1 \, d\tau_2 = \tfrac{1}{4} n(n+2) - S(S+1)$$

$$\text{Tr } \pi_t = \int \pi_t(\vec{r}_1, \vec{r}_2) \, d\tau_1 \, d\tau_2 = \tfrac{3}{4} n(n-2) + S(S+1) \tag{13}$$

Both π_s and π_t are non-negative functions, which means that either of them has to vanish if its trace happens to vanish. Consider two special cases:

1. $S = \tfrac{n}{2}$, then $\text{Tr } \pi_s = 0$, hence $\pi_s = 0$ and $\pi = \pi_t$.

2. $n = 2; S = 0$ then $\text{Tr } \pi_t = 0$, hence $\pi_t = 0$ and $\pi = \pi_s$.

These cases are especially interesting from the point of view of correlation.

38

B. Statistically Independent and Correlated Distributions

Consider two statistic variables, say x and y, for each of which a distribution function, say $f(x)$ and $g(x)$, and for both of which a joint distribution function, $F(x,y)$, is defined. We call the variables "independent" if and only if

$$F(x,y) = f(x)\,g(y) \qquad (14)$$

otherwise we call them "correlated". We here neglect the fact that a difference is sometimes made between independent and uncorrelated [14].

In order to apply the concepts of probability theory to electron correlation, we have to consider the case where the two variables have the same distribution function. We have also to take into account that $\varrho(\vec{r})$ is not the distribution function for one electron but for n electrons, in the same way as $\pi(\vec{r}_1, \vec{r}_2)$ is the joint distribution function for $n(n-1)$ pairs. The equation defining independent electrons therefore becomes [14]

$$\pi_{\text{ind}}(\vec{r}_1, \vec{r}_2) = \frac{n-1}{n}\,\varrho(\vec{r}_1)\,\varrho(\vec{r}_2) \qquad (15)$$

The factor $\frac{n-1}{n}$ is usually forgotten. The reason for this neglect is probably that electron correlation was first discussed for extended systems [1], where it is justifiable to neglect $\frac{1}{n}$ with respect to 1.

For finite systems it would be unreasonable to require Eq. (15) to have 1 instead of $\frac{n-1}{n}$ as the criterion for independence, because this equation could never be fulfilled for a simple reason of normalization. This can lead to the misleading conclusion that correlation is able to change the normalization.

C. The Causes of Electron Correlation

The Fermi Hole and the Coulomb Hole

If we take physical electron distribution and pair distribution functions, we realize that Eq. (15) is never fulfilled, but that in general $\pi(\vec{r}_1, \vec{r}_2)$ is smaller than predicted by (15) for small distances between $\vec{r}_1$ and $\vec{r}_2$ and larger in other regions of space. It is apparent that electrons try to avoid each other [2]. What are the reasons for this? There are actually three reasons for electron correlation in space.

1. The Pauli principle, *i.e.* the antisymmetry of the total wave function.

2. Certain spin and spatial symmetry requirements of the electron state.

3. The Coulomb repulsion between the electrons.

The first and the second reasons are related to each other. To understand their relation and their difference we have to go back to Wigner and Seitz's paper [1], where the Fermi correlation is defined. The wave function used there is the simplest one compatible with the Pauli principle, namely a single Slater determinant. Alternatively to the decomposition of π into π_s and π_t introduced in Section II. A, one can break down the functions ϱ and π in the following way [11,13]:

$$\varrho(\vec{r}) = \varrho^a(\vec{r}) + \varrho^\beta(\vec{r}) \tag{16}$$

$$\pi(\vec{r}_1, \vec{r}_2) = \pi^{aa}(\vec{r}_1, \vec{r}_2) + \pi^{a\beta}(\vec{r}_1, \vec{r}_2) + \pi^{\beta a}(\vec{r}_1, \vec{r}_2) + \pi^{\beta\beta}(\vec{r}_1, \vec{r}_2) \tag{17}$$

Here e.g., ϱ^a is the probability density to find an electron with α spin at the point $\vec{r}$ and $\pi^{a\beta}(\vec{r}_1, \vec{r}_2)$ is the probability density to find simultaneously one electron with α spin at $\vec{r}_1$ and a second electron with β spin at $\vec{r}_2$.

If the wave function from which ϱ and π are derived is a single Slater determinant, one can easily show that

$$\pi^{aa}(\vec{r}_1, \vec{r}_1) = 0 \tag{18}$$

$$\pi^{\beta\beta}(\vec{r}_1, \vec{r}_1) = 0 \tag{19}$$

$$\pi^{a\beta}(\vec{r}_1, \vec{r}_2) = \varrho^a(\vec{r}_1)\,\varrho^\beta(\vec{r}_2) \tag{20}$$

$$\pi^{\beta a}(\vec{r}_1, \vec{r}_2) = \varrho^\beta(\vec{r}_1)\,\varrho^a(\vec{r}_2) \tag{21}$$

This means that, if we describe a state by a single Slater determinant, we find *no* correlation between electrons with *different* spin. The probability of finding two electrons with the *same* spin at the same point $\vec{r}_1$ in space is zero. Any electron is surrounded by a "hole" into which no other electron with the same spin can penetrate. One can also show that the first derivative of $\pi^{aa}(\vec{r}_1, \vec{r}_2)$ with respect to $|\vec{r}_1 - \vec{r}_2|$ vanishes in all directions at $\vec{r}_1 = \vec{r}_2$. So if we plot $\pi^{aa}(\vec{r}_1, \vec{r}_2)$ schematically for fixed $\vec{r}_1$ as a function of $\vec{r}_2$ we get something like Fig. 2.

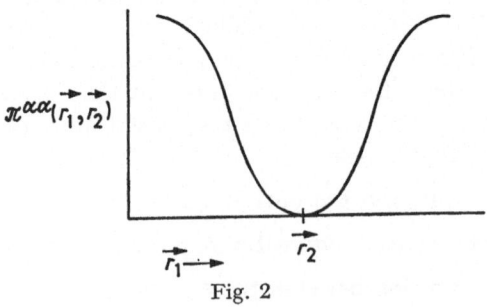

Fig. 2

This hole and this type of correlation is a direct consequence of the antisymmetry of the wave function. It has nothing to do with the repulsion between the electrons. It would still be there even if the electrons were attracting each other. The name "Fermi hole" for this type of situation goes back to Wigner and Seitz [1] and is now commonly used.

It is sometimes forgotten that, although Eqs. (18) and (19) hold for any wave function, Eqs. (20) and (21) are true only if the wave function is a Slater determinant. For physically realistic wave functions, electrons with different spin are correlated as well and in some cases this correlation can be of exactly the same type as the Fermi correlation. (As the definition of Fermi correlation states that it occurs only for electrons with the same spin, it must not be called Fermi correlation).

A simple example is a 1s2s configuration of a two-electron atom. If we say $1s = a$, $2s = b$, then neither of the two Slater determinants

$$\phi_1 = |a\alpha, b\beta|; \quad \phi_2 = |a\beta, b\alpha| \tag{22}$$

is a good description of the state of this configuration, since neither ϕ_1 nor ϕ_2 is an eigenfunction of S^2. These linear combinations

$$\Psi_1 = \frac{1}{\sqrt{2}}(\Phi_1 + \Phi_2) \quad \Psi_2 = \frac{1}{\sqrt{2}}(\Phi_1 - \Phi_2) \tag{23}$$

are eigenstates of S^2. Now Ψ_1, which belongs to the triplet state (with $M_s = 0$), must have the same ϱ and π as the corresponding triplet function (with $M_s = 1$).

$$\Psi_3 = |a\alpha, b\alpha| \tag{24}$$

(all spin-free quantities are the same for Ψ_1 and Ψ_3, but e.g. ϱ^a or $\pi^{a\beta}$ will be different for the two states).

For Ψ_1 it must hold that

$$\pi^{a\beta}(\vec{r}_1, \vec{r}_1) = \pi^{\beta a}(\vec{r}_1, \vec{r}_1) = 0. \tag{25}$$

This follows directly from the decomposition of π into π_s and π_t and the trace relationships (13). Note that for a triplet state $S = 1$, and that for a two-electron triplet state $n = 2$.

So we have a special case of $S = \frac{n}{2}$ which we now consider in detail.

The condition $S = \frac{n}{2}$ corresponds to the highest possible multiplicity compatible with the number of electrons. We have shown at the end of

Section II.A that $S = \frac{n}{2}$ implies that $\pi = \pi_t$. From Eq. (12), *i.e.* from the antisymmetry of π_t with respect to exchange of $\vec{r}_1$ and $\vec{r}_2$ (or $\vec{r}_1$ and $\vec{r}_2$), it follows that

$$\pi_t(\vec{r}_1, \vec{r}_1; \vec{r}_1, \vec{r}_1) = \pi(\vec{r}_1, \vec{r}_1) = 0. \tag{26}$$

This means that for a state with $S = \frac{n}{2}$ the probability of finding two electrons at the same point is zero. This result is independent of the spin of the respective electrons. Eq. (26) implies both (18/19) and (25).

Two electrons of different spin coupled to a triplet have the same kind of negative correlation as two electrons with the same spin. We call this kind of correlation "spin-induced correlation".

Now let us consider the spin-induced correlation of two electrons coupled to a singlet.

In order to understand this, let us again consider the ¹S and ³S terms of the 1s2s configuration of a two-electron atom. In the framework of Slater's theory of atoms the electron density $\varrho(\vec{r})$ is the same for both states, and for an exact treatment we would expect $\varrho(\vec{r})$ to differ very little between the two states. The pair density $\pi(\vec{r}_1, \vec{r}_2)$ is, however, quite different for the ¹S and the ³S states. If we call the two singly occupied real (orthogonal) orbitals a and b, we have (where ψ stands for the spin-free wave function and where a (1) has the same meaning as a ($\vec{r}_1$)

¹S: $\psi = \dfrac{1}{\sqrt{2}} \left[a(1)\, b(2) + b(1)\, a(2) \right]$

$\varrho(1) = |a(1)|^2 + |b(1)|^2$

$\pi(1,2) = |a(1)|^2\, |b(2)|^2 + |b(1)|^2\, |a(2)|^2 + 2\, a(1)\, b(1)\, a(2)\, b(2) \tag{27}$

³S: $\psi = \dfrac{1}{\sqrt{2}} \left[a(1)\, b(2) - b(1)\, a(2) \right]$

$\varrho(1) = |a(1)|^2 + |b(1)|^2$

$\pi(1,2) = |a(1)|^2\, |b(2)|^2 + |b(1)|^2\, |a(2)|^2$

$$- 2\, a(1)\, b(1)\, a(2)\, b(2) \tag{28}$$

(Of course, (28) is true for any of the three components with $M_s = 1, 0, -1$).

We see immediately that for the ³S state $\pi(1,1) = 0$, but for the ¹S state we have

$$\pi(1,1) = 4|a(1)|^2\, |b(1)|^2 \tag{29}$$

The result for the independent distributions according to Eq. (15) would be

$$\pi_{ind}(1,1) = \tfrac{1}{2}\,\varrho(1)\,\varrho(1) = \tfrac{1}{2}\,\{|a(1)|^2 + |b(1)|^2\}^2$$

$$= |a(1)|^2\,|b(1)|^2 + \tfrac{1}{2}\,|a(1)|^4 + \tfrac{1}{2}\,|b(1)|^4 \tag{30}$$

Now let us take a point in space for which $|a(1)| = |b(1)|$. For such a point $\pi(1,1)$ of (29) is twice as large as $\pi_{ind}(1,1)$ of (30), which means that the probability of finding the two electrons there simultaneously is twice as high as we would expect for independent events. The electron correlation is positive rather than negative as for the triplet state.

Here there is a correlation hump rather than a correlation hole. The electrons do not avoid each other, but they approach closer to each other than independent electrons would.

At first glance it seems surprising that there is a possibility of either negative or positive correlation, although the electrons repel each other and negative correlation is energetically more favorable. But this energetic preference for negative correlation is apparent in the fact that, of the two states of the same configuration, the one with the more negative correlation, namely the ^3S state, has the lower energy, which is in agreement with Hund's rule. In fact, Hund's rule is nothing else but the statement that, among states with the same or approximately the same $\varrho(\vec{r})$, the one with the most negative correlation — i.e. the one which best allows the electrons to avoid each other — has the lowest energy. States with positive electron correlation can exist, but are normally not ground states. If we wish to apply the same reasoning to, say, the ^3P, ^1D and ^1S state of a p^2 configuration i.e. to describe the "symmetry-induced correlation", we must note that only the totally invariant (i.e. rationally symmetric) parts [13], ϱ^0 and π^0, affect the energy and that the three states have the same ϱ^0 but different π^0.

Finally we have to discuss the direct consequences of electron repulsion for the correlation of the electrons in space. We consider a state where neither Fermi correlation, nor spin nor symmetry-induced correlation is possible, namely the ground state of a two-electron system like He or H_2. The simplest possible approximate wave function for such a state is of the form (spin-free)

$$\psi(1,2) = \phi(1)\,\phi(2) \tag{31}$$

It is easily appreciated that for this particular wave function Eq. (15) is verified and hence, if (31) were a solution of the Schrödinger equation, there would be no correlation at all.

Kato [15] has derived very interesting theorems that hold for the true solutions of the Schrödinger equation, *e.g.* for a two-electron system ($\overline{\psi}$ means the angular average of ψ over the angle θ between $\vec{r}_1$ and $\vec{r}_2$)

$$\lim_{\vec{r}_2 \to \vec{r}_1} \frac{\partial \overline{\psi}\,(\vec{r}_1, \vec{r}_2)}{\partial r_{12}} = \tfrac{1}{2}\,\psi(\vec{r}_1, \vec{r}_1) \tag{32}$$

This means that the first derivative of the angular average of the wave function with respect to the interelectronic distance at points where the positions of the two electrons coincide is related to the value of the wave function at the same points. If the latter value does not happen to vanish, then the first derivative is non-zero as well. This means that $\psi(\vec{r}_1, \vec{r}_2)$ has a cusp (*i.e.* a discontinuity in its first derivative) as illustrated in Fig. 3.

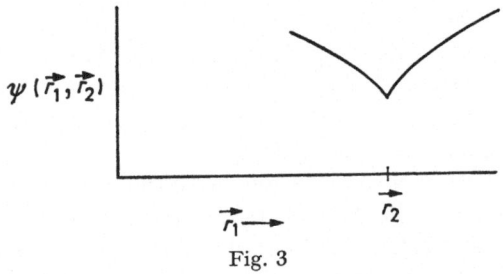

Fig. 3

Bingel [16] has investigated the consequence of Kato's theorems for the pair density. It seems that generally the pair density has a cusp like that shown in Fig. 3. The Coulomb hole is not as deep as the Fermi hole and it has a cusp.

It is possible to expand the true wave function in its canonical [17] or natural [18] form.

$$\psi(1,2) = \sum_k c_k\,\phi_k\,(1)\,\phi_k(2) \tag{33}$$

Although we can obtain rather good approximations to the true wave function by using *finite* expansions of the form (33), very high accuracy is difficult to achieve. The reason for this is understandable. Since the ϕ_k in (33) are (usually) analytic functions everywhere (except possibly at the positions of the nuclei), $\psi(1,2)$ being a finite expansion is also analytic, in particular at those points in configuration space where $\vec{r}_1 = \vec{r}_2$. So this $\psi(1,2)$ cannot have a cusp. We are thus faced with the problem of representing a function with a discontinuous first derivative as the limit of a sequence of analytic functions. Such sequences converge very slowly, if at all. The problem is in fact very similar to that of reproducing the cusp of the electron density at the nuclei by a linear combination of functions that are differentiable at the positions of the nuclei (*e.g.* one-centre expansion of H_2, use of Gaussian wave functions etc.).

It has been plausibly argued [19-20] that for the He atom the energy converges as follows:

$$E = \sum_{l=0}^{\infty} \varepsilon_l, \quad \varepsilon_l \sim (l + \tfrac{1}{2})^{-4} \text{ for large } l \tag{34}$$

where ε_l is the contribution to the energy of orbitals in (33) that have the angular quantum number l. For triplet states of two-electron systems $\psi(\vec{r}_1, \vec{r}_1)$ vanishes and there is no correlation-cusp problem (at least not of the same kind as for singlet states) so that the convergence of the natural expansion [18]

$$\psi(1,2) = \sum_{k} a_k \left[u_k(1) \, v_k(2) - v_k(1) \, u_k(2) \right] \tag{35}$$

is usually much more rapid.

D. Left-Right, in-out, and Angular Correlation

We now discuss the correlation of the electrons in space for the ground state of the H_2 molecule[a], writing its wave function in the natural expansion form. Let us first limit ourselves to a two-term expansion:

$$\psi(1,2) = c_1 \, \sigma_g(1) \, \sigma_g(2) + c_2 \, \sigma_u(1) \, \sigma_u(2) \tag{36}$$

If we want to calculate the best two-term expansions for the H_2 ground state, we find in fact that ϕ_1 is of σ_g and ϕ_2 of σ_u type. We also find that $c_1 \approx 0.99$, $c_2 \approx -0.1$.

The electron density and the pair density corresponding to (36) are

$$\begin{aligned}
\varrho(1) &= 2 \, c_1^2 \, |\sigma_g(1)|^2 + 2 \, c_2^2 \, |\sigma_u(1)|^2 \\
\pi(1,2) &= 2 \, c_1^2 |\sigma_g(1)|^2 \, |\sigma_g(2)|^2 + 2 \, c_2^2 \, |\sigma_u(1)|^2 \, |\sigma_u(2)|^2 \\
&\quad + 4 \, c_1 \, c_2 \, \sigma_g(1) \, \sigma_u(2) \, \sigma_u(1) \, \sigma_g(2)
\end{aligned} \tag{37}$$

We now use the fact that $|c_2| << |c_1|$; we therefore neglect c_2^2 with respect to c_1^2 or $c_1 c_2$.

Since $c_1^2 = 1 - c_2^2$ we can then also put $c_1 = 1$. So we get (except for terms of the order c_2^2):

$$\begin{aligned}
\varrho(1) &\approx 2 \, |\sigma_g(1)|^2 \\
\pi(1,2) &\approx 2 \, |\sigma_g(1)|^2 \, |\sigma_g(2)|^2 + 4 \, c_2 \, \sigma_g(1) \, \sigma_u(1) \, \sigma_g(2) \, \sigma_u(2) \\
&= \tfrac{1}{2} \varrho(1) \, \varrho(2) + 4 \, c_2 \, \sigma_g(1) \, \sigma_u(1) \, \sigma_g(2) \, \sigma_u(2)
\end{aligned} \tag{38}$$

---

[a] This correlation is a purely Coulomb correlation.

If the terms with the factor $4\,c_2$ were zero we could fulfil Eq. (15). This term thus contains the kind of electron correlation introduced by our ansatz. In order to get a pictorial idea of this type of correlation, we introduce as further simplification the LCAO approximation.

$$\sigma_g = \frac{1}{\sqrt{2(1+S)}}\,[a(1) + b(1)]$$

$$\sigma_u = \frac{1}{\sqrt{2(1-S)}}\,[a(1) - b(2)] \tag{39}$$

where a and b are atomic orbitals localized on either nucleus. Then we get

$$\pi(1,2) - \tfrac{1}{2}\,\varrho(1)\,\varrho(2) = 4\,c_2\,\sigma_g(1)\,\sigma_u(1)\,\sigma_g(2)\,\sigma_u(2) \tag{40}$$

$$= \frac{c_2}{1-S^2}\left\{ |a(1)|^2\,|a(2)|^2 + |b(1)|^2\,|b(2)|^2 - |a(1)|^2\,|b(2)|^2 - |b(1)|^2\,|a(2)|^2 \right\}$$

We see that the probability of finding one electron in the AO a and the other in the AO b is increased (c_2 is negative) by $\frac{c_2}{1-S^2}$, while the probability of finding the two electrons in the same AO is reduced. The admixture of $\sigma_u(1)\,\sigma_u(2)$ to the leading configuration $\sigma_g(1)\,\sigma_g(2)$ allows the electrons to avoid each other, so that, if the one is in the AO a, the other is somewhat more likely to be in the AO b than it would be in the case of independent electrons. If one electron is on the left of the molecule the other is more likely to be on the right, therefore this type of correlation is often called "left-right correlation".

The next important natural orbitals of the H_2 ground state are of the type π_u, $\bar{\pi}_u$ and $2\sigma_g$. A wave function containing these terms yields 99.5% of the total energy and 98% of the binding energy of H_2. If one analyzes the role of these additional configurations in a similar pictorial way, one finds that the admixture of $\pi_u(1)\,\bar{\pi}_u(2)$ and $\bar{\pi}_u(1)\,\pi_u(2)$ allows for what is called "angular correlation". The second electron has some preference for a position diametral to the first with respect to the internuclear axis. The $2\sigma_g$ orbital allows for "in-out correlation": If one electron is close to the internuclear axis, the other tends to be distant from it.

The higher NO's may be given similar though more complicated pictorial interpretations, but their role is rather to represent — as far as this is possible — the correlation cusp. Usually very few configurations convey the bulk effect of electron correlation and very many are needed to get the remaining few per cent.

We note that the improvement of a wave function by configuration interaction leads mainly to a change in the two-particle density $\pi(1,2)$, not in the electron density $\varrho(1)$: this allows the electrons to avoid each other without much changing the total electron distribution. This is due to the fact that the coefficients c_2, c_3 etc. in the natural expansion are small in absolute value and that they enter π linearly but ϱ only quadratically. One can estimate the c_k by perturbation theory [22]

$$c_k \approx \frac{(1k \mid k1)}{E_1 - E_k} \tag{41}$$

where $(1k \mid k1)$ is an exchange integral between ϕ_1 and ϕ_k (which is always positive) and where E_1 and E_k are the one-configuration energy expectation values of $\phi_1(1)\,\phi_1(2)$ and $\phi_k(1)\,\phi_k(2)$, respectively. Exchange integrals are usually small compared to the energy differences in the denominator. This is the normal case that one refers to as "dynamic correlation".

The situation is somewhat different if the denominator happens to be small, *i.e.* if there is some configuration that is degenerate or near degenerate with the one that we want to improve by configuration interaction. Then the corresponding c_k may become rather large in absolute value. This occurs whenever there is a crossing of potential curves for single determinant wave functions. This situation is sometimes referred to as "first-order configuration interaction" or "static correlation". In such a case even ϱ may be changed considerably by the configuration interaction.

III. The Calculation of Correlation Effects

A. Methods Based on a Direct Calculation of the Two-Particle Density Matrix

By means of Eqs. (5) and (6) the expectation values of one- and two-particle operators are expressed through the one- and two-particle density matrices; hence the energy of an atom or a molecule can be written as

$$E = \tfrac{1}{2} \operatorname{Tr}\{\overline{H}(1,2)\,\pi\,(1,2;\,1',\,2')\} \tag{42}$$

where

$$\overline{H}\,(1,\,2) = \frac{h(1) + h(2)}{n - 1} + \frac{1}{r_{12}} \tag{43}$$

is the so-called "reduced" or "Bopp" Hamiltonian and where Tr ($=$ trace) means that one must first apply $\overline{H}$ to the unprimed coordinates of π, then make the unprimed equal to the primed coordinates and integrate over $d\tau_1$ and $d\tau_2$.

47

Bopp[23] had the ingenious idea to determine $\pi(1,2; 1' 2')$ such that the energy expression (42) would be stationary with respect to variations of π. This led to the condition that this particular π is of the form

$$\pi(1,2; 1' 2') = 2 \sum_k \mu_k \; \omega_k \; (1,2) \; \omega_k(1' ,2') \tag{44}$$

where the $\omega(1,2)$ are eigenfunctions of $\bar{H}$ (1,2)

$$\bar{H} \; \omega_k(1,2) = \lambda_k \; \omega_k \; (1,2) \tag{45}$$

The minimum of the energy expression (42) is obtained if the ω_k with the lowest λ_k are filled up to their maximum allowed occupation number μ_k. Bopp [23] gave an erroneous proof that the maximum occupation number of any ω_k is one, *i.e.* that

$$0 \leq \mu_k \leq 1 \tag{46}$$

If (46) were true, then the ω_k with the $n(n{-}1)$ lowest λ_k should be $\mu_k = 1$, and all other ω_k should have $\mu_k = 0$.

By this approach the solution of the n-particle Schrödinger equation is reduced to that of a two-particle Schrödinger equation, which simplifies the problem considerably. Unfortunately there are two very serious objections that invalidate this approach completely

1. (46) is not true. The least upper bound [24,25,26] for any μ_k is $\left[\frac{n}{2}\right]$ rather than 1.

$$0 \leq \mu_k \leq \left[\tfrac{n}{2}\right] = \begin{cases} \dfrac{n}{2} \text{ for } n \text{ even} \\[2mm] \dfrac{n+1}{2} \text{ for } n \text{ odd} \end{cases} \tag{47}$$

2. Varying $\pi(1,2; 1' 2')$ free from any constraint violates the Pauli-Principle. In other words, not any arbitrary π that one can write down is derivable (through Eq. (8)) from an anti-symmetric n-electron wave function [27,28]. The problem, to find sufficient and necessary conditions for π to be derivable from an antisymmetric n-electron Ψ (without explicitly referring to the particular Ψ) or, in other words, for π to be n-representable, has been called the n-representability problem [26].

Much effort has been expended in trying to solve this n-representability problem. Many necessary conditions have been formulated, *e.g.* behaviour with respect to particle exchange (12), relation to total spin (13), bounds on the eigenvalues (46), and many others [26,29,30]. Unfortunately no complete set of simple necessary conditions has been found that is also sufficient.

One can, of course, modify Bopp's approximation and impose certain known necessary n-representability conditions as constraints to the variations of π in Eq. (42). Some attempts have been made along these lines [31,32] but — except possibly for extended systems — they have so far not been too promising. Another possibility is [33] not to require that π is exactly n-representable but that it is "approximately n-representable" or "n-representable in some limit". Take as an example a series of isoelectronic atoms. In the limit $\frac{1}{Z} \to 0$ the electron interaction can be neglected, Ψ_0 can be calculated easily, and π_0 is obtained from Ψ_0. In a variational treatment based on Eq. (42) for finite Z one allows only such a functional form of π that for $Z \to \infty$ π converges towards the known π_0. Then one is sure that the π that one obtains is at least n-representable in the limit $Z \to \infty$.

It can be shown [33] but we are not going to do so here — that starting from the idea of "n-representability in the limit" one can derive a theory of independent electron pairs that is equivalent to the independent electron-pair approximation (IEPA) which we shall now derive in a more conventional way starting from a CI expansion.

B. Configuration Interaction and Definition of Pair Energies

We consider a quantum mechanical state which can to a good approximation be described by a single Slater determinant Φ built up from orthonormal spin orbitals $\psi_i (i = 1, 2 \ldots n)$. We complement the set of ψ_i by other spin orbitals ψ_a, $(a = n+1, \ldots n+m)$ such that the ψ_i together with the ψ_a, form an orthonormal set, which becomes complete in the limit $m \to \infty$.

We define "singly substituted" determinants Φ_i^a as those that are obtained from Φ if one replaces ψ_i by ψ_a. "Doubly" and higher substituted determinants Φ_{ij}^{ab}, Φ_{ijk}^{abc} etc. are similarly defined. The configuration-interaction (CI) expansion of the best wave function for the considered state can be written in terms of our Slater determinants as [34]:

$$\Psi = c_0\, \Phi + \sum_{i,a} c_i^a\, \Phi_i^a + \sum_{i<j} \sum_{a<b} c_{ij}^{ab}\, \Phi_{ij}^{ab} + \ldots \tag{48}$$

where the c_0, c_i^a etc. are coefficients and where both Ψ and any of the Φ, Φ_i^a etc. are normalized to unity. If the spin-orbital basis is complete, an expansion of the true Ψ in the form of Eq. (48) is possible.

We assume for the moment that we know the true wave function (if the spin-orbital basis is complete) or the best variational function in the given finite basis in the form of Eq. (48). Then it holds that

$$H\Psi = E\Psi \tag{49}$$

49

$$(\varPhi, H\varPsi) = E(\varPhi, \varPsi) = E\, c_0 \tag{50}$$

$$E = (\varPhi, H\varPhi) + \frac{1}{c_0}\sum_{i,a} c_i^a\, (\varPhi, H\varPhi_i^a) + \frac{1}{c_0}\sum_{i<j}\sum_{a<b} c_{ij}^{ba}(\varPhi, H\varPhi_{ij}^{ab}) + \cdots \tag{51}$$

where H is the Hamiltonian (if the basis is complete) or the projection of the Hamiltonian onto the subspace spanned by our determinants. We shall not distinguish formally between the two possibilities.

Matrix elements between $\varPhi$ and triply or higher substituted determinants vanish because H is a sum of one- and two-electron operators only, so in (51) the sign $+$ and the points at the end can be omitted.

The contribution

$$\frac{1}{c_0}\sum_{i,a} c_i^a\,(\varPhi, H\varPhi_i^a) \tag{52}$$

of the singly substituted configurations can be made to vanish by a proper unitary transformation of the spin-orbital basis $\psi_i,\ \psi_a$. There are in fact two possibilities:

1. to choose the ψ_i such that they are eigenfunctions of a (unrestricted) Hartree-Fock equation. Then the Brillouin Theorem is satisfied

$$(\varPhi, H\varPhi_i^a) = 0 \text{ for all } i,\ a \tag{53}$$

and (52) vanishes in fact.

2. to choose ψ_i such that, rather than (53), one has

$$c_i^a = 0 \text{ for all } i,\ a \tag{54}$$

This is the case if the ψ_i are chosen as the "best overlap" [35] or "Brueckner" [34,36] spin orbitals. That such a choice is always possible has been shown by Brenig [37] and independently by Nesbet [34]. Eq. (54) is usually referred to as the Brueckner condition, in contrast to the Brillouin condition (53). Note that (53) can be regarded as either a theorem, if one defines the Hartree-Fock equation in the conventional way, or as a condition from which the conventional Hartree-Fock equation can be derived.

If either (53) or (54) is fulfilled, the energy expression (51) becomes

$$E = E_0 + \sum_{i<j} \varepsilon_{ij} = E_0 + E_{\text{corr}} \tag{55}$$

with

$$E_0 = (\varPhi, H\varPhi) \tag{56}$$

$$\varepsilon_{ij} = \frac{1}{c_0}\sum_{a<b} c_{ij}^{ab}\,(\varPhi, H\varPhi_{ij}^{ab}) \tag{57}$$

We call E_{corr} the correlation energy and ε_{ij} the contribution of the spin orbital pair (ψ_i, ψ_j) to the correlation energy. With this definition the correlation energy can be expressed *exactly* as a sum of pair contributions. So far we have only been concerned with the definition of the ε_{ij} and we have not learned anything about how to calculate the ε_{ij}. Still in the same philosophy, we introduce some other convenient definitions.

Let us abbreviate a two-electron Slater determinant in the following way

$$[i, j] = \tfrac{1}{\sqrt{2}} [\psi_i(1) \, \psi_j(2) - \psi_i(2) \, \psi_j(1)] \tag{58}$$

and define the pair correction functions

$$u_{ij}(1,2) = \tfrac{1}{c_0} \sum_{a<b} c_{ij}^{ab} [a, b] \tag{59}$$

Then we can express the ε_{ij} of (57) as

$$\varepsilon_{ij} = <u_{ij}|h(1) + h(2) + \frac{1}{r_{12}} \mid [i, j]> \; = \; <u_{ij}| \frac{1}{r_{12}} \mid [i, j]> \tag{60}$$

Hence knowledge of the u_{ij} implies knowledge of the ε_{ij}, and knowledge of the u_{ij} and of c_0 is equivalent to knowledge of the c_{ij}^{ab}. If we find a way to calculate the pair correction functions u_{ij}, we have practically all necessary information about our system — at least as far as the energy is concerned. So the main problem is to calculate good approximations to the u_{ij}. We have to stress that Eq. (51) cannot be the starting point for a variational calculation of the u_{ij}, via the c_{ij}^{ab}, because (51) is *not* an upper bound to the true energy. The variation of (51) does not vanish if the c_{ij}^{ab} are the exact ones.

Note that the definitions in Eq. (55—57) depend on whether we have chosen the Brillouin (53) or the Brueckner condition (54), so we should, when necessary, state this choice explicitly. Numerically the differences are usually very small.

C. Calculation of the Correlation Energy from Approximate Pair-Correlation Functions

We are still postponing the question of how to calculate the pair-correction functions u_{ij}, but let us assume that somehow we have got approximations $\tilde{u}_{ij}$ to them. If we use these $\tilde{u}_{ij}$ and insert them into (60) we get approximate pair-correlation energies $\tilde{\varepsilon}_{ij}$. One can usually presume — since $\frac{1}{r_{12}}$ is a relatively bounded operator with respect to the Hamiltonian — that if $\tilde{u}_{ij}$ is sufficiently close to u_{ij} then $\tilde{\varepsilon}_{ij}$ too will be sufficiently close to ε_{ij}. However, $\sum_{i<j} \tilde{\varepsilon}_{ij}$ is *not* an upper bound to E_{corr} so that on improving the $\tilde{u}_{ij}$ one does *not* generally approach the ε_{ij} from above, as one is accustomed to do in genuine variational calculations.

One nice thing about variational calculations is that the difference with respect to the true energy is a direct measure and a rather severe test of the quality of the calculation. It is justifiable to keep the intrinsic test for the calculations even at the expense of much labour, and to be sceptical with respect to calculations that are not genuinely variational. On the other hand, one should not overestimate the value of having an upper bound to the energy without any theoretical criterion as to how close this upper bound is to the true energy (if the latter happens to be unknown).

It is possible, though sometimes disregarded, to calculate an approximation to the correlation energy that *is* an upper bound to the true E_{corr} from approximate u_{ij}'s. This is done in the following way.

Starting from the $\tilde{u}_{ij}$'s one immediately gets the $\dfrac{c_{ij}^{ab}}{c_0}$, one can then write down the wave function (48) limited to doubly substituted determinants

$$\Psi_{(2)} = c_0' \Phi + \sum_{i<j} \sum_{a<b} \frac{c_{ij}^{ab}}{c_0} \, \Phi_{ij}^{ab} \tag{61}$$

Remember that we suppose the $\dfrac{c_{ij}^{ab}}{c_0}$ to be known rather than the c_{ij}^{ab} and c_0 separately. The constant c_0' is determined by the requirement that $\Psi_{(2)}$ should be normalized to unity. There is no reason why c_0 and c_0' should be identical, though they are usually not very different. Since we know $\Psi_{(2)}$, we can without particular difficulty calculate the expectation value

$$E_{(2)} = (\Psi_{(2)}, H\Psi_{(2)}) \tag{62}$$

So one can derive two different energy expressions from the $\tilde{u}_{ij}$, either

$$\tilde{E} = E_0 + \sum_{i<j} \tilde{\varepsilon}_{ij} = E_0 + \sum_{i<j} <\tilde{u}_{ij} \left| \frac{1}{r_{12}} \right| [i, j]> \text{ or } E_{(2)}. \tag{63}$$

Which of the two energies is better, *i.e.* closer to the true energy E? If $\tilde{u}_{ij}$ is a crude approximation to $\tilde{u}_{ij}$, then $E_{(2)}$ inspires somewhat more confidence since it is an upper bound to E. If, on the other hand we happen to know the exact u_{ij} (at least the best variational u_{ij} within our basis) then $\tilde{E}$ of (63) which becomes E of (55) is *exact*, whereas $E_{(2)}$ of (62) is *not*. In fact, in order to calculate the exact E as an expectation value, we need more information about Ψ than just the knowledge of the $\dfrac{1}{c_0} c_{ij}^{ab}$. Including other than doubly substituted determinants in Ψ can — in the sense of the variation principle — only lower the energy. In other words, if we calculate

$E_{(2)}$ from (62) using the exact u_{ij}, we get an energy that is above the exact energy in proportion to the extent of the contributions of the ϕ_{ijk}^{abc} etc. to Ψ.

If we find that $\tilde{E}$ is lower than $E_{(2)}$, this does not necessarily mean that this difference is a result of the poorness of the $\tilde{u}_{ij}$; it may also mean that the $\tilde{u}_{ij}$ are quite good and that $\tilde{E}$ is a better approximation to E than is $E_{(2)}$. Since $\tilde{E}$ is calculated much more easily than $E_{(2)}$ there is a temptation to forget about $E_{(2)}$ and to regard $\tilde{E}$ as *the* approximate energy.

One way to improve $E_{(2)}$ is to include in Ψ also the socalled unlinked clusters [37-40, 67)] of pair substitutions. We come back to this point.

D. The APSG Ansatz

We come now to the problem of how to calculate the pair-correction functions u_{ij}. The oldest approach is the one first introduced by Hurley, Lennard-Jones and Pople [17)]. These authors started from a very limited ansatz rather than the general CI expansion of Eq. (48). We formulate this ansatz in the following way to point out its relation to the general pair theory. We limit ourselves to a system with an even number n of electrons and build up Φ from the spin orbitals $\phi_R\alpha$ and $\phi_R\beta$ with $R = 1, 2 \ldots \frac{n}{2}$.

Let us partition the Hilbert space spanned by the ψ_i, ψ_a into $\frac{n}{2}$ orthogonal subspaces. Each subspace is associated with one of the ϕ_R that are doubly occupied in Φ. We can choose an orthogonal basis in any of these subspaces and label these basis orbitals as ϕ_k^R. We also define $\phi_1^R = \phi_R$.

Then the orthogonality of the subspace implies that

$$(\phi_k^R, \phi_l^S) = \delta_{RS} \delta_{kl} \tag{64}$$

In the CI expansion (48) we only include those doubly substituted Slater determinants $\Phi_{R\bar{R}}^{k\bar{l}}$ that are obtained from Φ if one replaces $\phi_R\alpha$ and $\phi_R\beta$ by $\phi_k^R\alpha$ and $\phi_l^R\beta$. We include as well quadruply substituted determinants of the form $\Phi_{RRSS}^{k\bar{l}m\bar{n}}$, *i.e.* obtained from Φ on substituting $\phi_R\alpha$, $\phi_R\beta$, $\phi_S\alpha$, $\phi_S\beta$ by $\phi_k^R\alpha$, $\phi_l^R\beta$, $\phi_m^S\alpha$, $\phi_n^S\beta$ and 6-fold substituted determinants $\Phi_{RRSSTT}^{k\bar{l}m\bar{n}p\bar{q}}$ etc. We do not, however regard the coefficients of these higher than doubly substituted determinants as free variational parameters, but we make a

53

condition that these coefficients are expressed through those of the doubly substituted determinants like

$$c_0 \cdot c_{R\bar{R}S\bar{S}}^{k\,\bar{l}m\bar{n}} = c_{R\bar{R}}^{k\,\bar{l}} \cdot c_{S\bar{S}}^{m\bar{n}} \tag{65}$$

$$c_0^2 \cdot c_{R\bar{R}S\bar{S}T\bar{T}}^{k\,\bar{l}m\bar{n}p\bar{q}} = c_{R\bar{R}}^{k\,\bar{l}} \cdot c_{S\bar{S}}^{m\bar{n}} \cdot c_{T\bar{T}}^{p\bar{q}}$$

Thus we have the $c_{R\bar{R}}^{k\,\bar{l}}$ as the only variational parameters. Fourfold, six-fold etc. substitutions are included only in as far as they are "products" of double substitutions; 2 n-fold substitutions of this particular form are referred to as "unlinked clusters".

For this ansatz the pair-correction functions $u_R(1,2)$ are given as

$$u_R(1,2) = \frac{1}{c_0} \sum_{\substack{k,l \\ (>0)}} c_{R\bar{R}}^{k\,\bar{l}} \, [R^k\alpha, R^l\beta] \tag{66}$$

(there are only $\frac{n}{2}$ of them instead of $\frac{n(n-1)}{2}$ as in the general case). The expansion of the wave function is

$$\Psi = c_0\Phi + \sum_R \sum_{k,l} c_{R\bar{R}}^{k\,\bar{l}} \, \Phi_{R\bar{R}}^{k\,\bar{l}} + \sum_{R<S} \sum_{k,l,m,n} c_{R\bar{R}}^{k\,\bar{l}} \, c_{S\bar{S}}^{m\bar{n}} \, \Phi_{R\bar{R}S\bar{S}}^{k\,\bar{l}m\bar{n}} + \cdots \tag{67}$$

If we introduce the normalized pair function or geminal

$$\omega_R(1,2) = N_R\,[R^1\alpha, R^1\beta] + u_R(1,2) \tag{68}$$

we can write Ψ in the form

$$\Psi = A\left\{\prod_{R=1}^{\frac{n}{2}} \omega_R\,(2R-1, 2R)\right\} \tag{69}$$

The normalization factors N_R of the geminals ω_R are related to the coefficient c_0 in (67) through

$$c_0 = \prod_{R=1}^{\frac{n}{2}} N_R \tag{70}$$

Since our wave function is most conveniently written in the form (69), *i.e.* as an antisymmetrized product of strongly orthogonal geminals [41], the name APSG is now commonly used for this ansatz.

The expectation value $<H>$ of the Hamiltonian for an APSG wave function assumes a rather simple form, in particular if one expresses the ω_R in terms of their natural expansion, *i.e.* if one chooses the functions ϕ_R^k such that ω_R is diagonal (which involves no loss in generality).

$$\omega_R(1,2) = \sum_{k=1}^{\infty} d_k^R \phi_k^R(1) \phi_k^R(2) \, {}^1\theta(1,2) \tag{71}$$

where ${}^1\theta$ is the singlet spin function.

It is possible to base on the APSG ansatz a rigorous variational procedure, *i.e.* to minimize $<H>$ with respect to the ω_k, subject to their strong orthogonality which is equivalent to Eq. (64), and to calculate the optimum ω_k without introducing further simplifications [41]. Applications to some small molecules by K. Ruedenberg *et al.* [42,43] have to be mentioned in this context. This is a substantial advantage of this ansatz, but there are also some serious drawbacks.

1. The percentage of the correlation energy accounted for is sufficiently large ($> 95\%$) for systems [42,44] like Be or LiH, whereas it becomes rather poor (~ 40–50% only) for molecules like BH, CH_4 etc.[b].

2. The equation to be solved in order to get the optimum geminals are coupled in a rather complicated way and their solution requires much computer time.

Before we discuss how to overcome these drawbacks, we consider two important points in connection with the APSG scheme [44].

1. The relation to an independent electron-pair approximation.

2. The role of the unlinked clusters.

To do so we compare the APSG wave function (67) with a wave function Ψ_R in which only the spin-orbital pair $\phi_R\alpha\phi_R\beta$ is substituted, but with the substitution limited to the subspace of Hilbert space associated with ϕ_R. The wave function Ψ_R can then also be written in the APSG form (69), where the geminals ω_S for $S \neq R$ are Slater determinants built up from the spin orbitals $\phi_S\alpha$, $\phi_S\beta$. It is convenient to expand the geminals in their natural form (71). In terms of the natural orbitals Ψ_{APSG} and Ψ_R have the following CI expansions respectively:

$$\Psi_{\text{APSG}} = c_0\Phi + \sum_R \sum_{\substack{k \\ (>1)}} d_k^R \Phi_{R\bar{R}}^{k\bar{k}} + \sum_{R<S} \sum_{k<l} d_k^R d_l^S \Phi_{RRSS}^{k\bar{k}l\bar{l}} + \cdots \tag{72}$$

---

[b] The poor result for CH_4 is mainly due to the importance of the interpair correlation, and that for BeH to the orthogonality constraint (see Section IV and Ref. [63]).

$$\Psi_R = c_1^R \Phi \;+\; \sum_{\substack{k \\ (>1)}} d_k^R \, \Phi_{R\bar{R}}^{k\bar{k}} \tag{73}$$

The normalization condition for the geminal (71) is

$$\sum_{k=1}^{\infty} |d_k^R|^2 = 1 \tag{74}$$

With this normalization Ψ_R is automatically normalized to unity and Ψ_{APSG} is so if

$$c_0 = \prod_R d_1^R \tag{75}$$

This is only true if we keep the Hilbert space factorized, so that (64) holds.

If we abbreviate the diagonal matrix elements of the one-electron part of the Hamiltonian with respect to ϕ_R^R as H_{kk}^R and use the Mulliken notation for two-electron integrals, we get the following expectation values E_{APSG} for Ψ_{APSG}, E_R for Ψ_R and E_0 for the single Slater determinant. In all cases we limit ourselves to a closed-shell state and we assume the coefficients to be real.

$$
\begin{aligned}
E_{\text{APSG}} = 2 \sum_{R,i} (d_i^R)^2 \, H_{ii}^R + \sum_{R,i,j} d_j^R \, d_j^R (R_i R_j | R_j R_i) \\
+ \sum_{R,S}' \sum_{i,j} (d_i^R)^2 \, (d_j^S)^2 \, \{2(R_i R_i | S_j S_j) - (S_j R_i | R_i S_j)\}
\end{aligned}
\tag{76}
$$

$$
\begin{aligned}
E_R = 2\sum_i (d_i^R)^2 \, H_{ii}^R + 2\sum_{\substack{S \\ (\neq R)}} H_{11}^S + \sum_{i,j} d_i^R \, d_j^R \, (R_i R_j | R_j R_i) \\
+ 2 \sum_i \sum_{\substack{S \\ (\neq R)}} (d_i^R)^2 \, \{2 \, (R_i R_i | S_1 S_1) - (S_1 R_i | R_i S_1)\}
\end{aligned}
\tag{77}
$$

$$
\begin{aligned}
E_0 = 2\sum_R H_{11}^R + \sum_{R,S}' \, \{2 \, (R_1 R_1 | S_1 S_1) - (S_1 R_1 | R_1 S_1)\} \\
+ \sum_R (R_1 R_1 | R_1 R_1)
\end{aligned}
\tag{78}
$$

The correlation energy, as accounted for by Ψ_{APSG} and Ψ_R respectively, is defined as follows:

$$E_{\text{corr}}^{\text{APSG}} = E_{\text{APSG}} - E_0 \tag{79}$$

$$E_{\text{corr}}^R = E_R - E_0 \tag{80}$$

The following relation between $E_{\text{corr}}^{\text{APSG}}$ and the E_{corr}^R is easily found

$$E_{\text{corr}}^{\text{APSG}} = \sum_R E_{\text{corr}}^R + {\sum_{R,S}}' \sum_{i,j} a_i^R a_j^S \{2\,(R_i\,R_i|S_j\,S_j) - (R_i\,S_j|S_j\,R_i)\} \tag{81}$$

with

$$a_i^R = \begin{cases} (d_i^R)^2 & \text{for } i > 1 \\[2mm] (d_1^R)^2 - 1 & \text{for } i = 1 \end{cases} \tag{82}$$

The individual terms in the quadruple sum in (81) are of fourth order in the small coefficients $d_i^R (i \neq 1)$ and they have different signs so that they even cancel to a large extent. If the expression in braces were independent of i and j, the quadruple sum would vanish exactly because of (74). In all practical cases the quadruple sum is of the order of $1^0/_{00}$ of $E_{\text{corr}}^{\text{APSG}}$ or less and is therefore negligible. This means

$$E_{\text{corr}}^{\text{APSG}} \approx \sum_R E_{\text{corr}}^R \tag{83}$$

Since E_{corr}^R is the correlation energy of the R-th pair calculated independently from the other pairs (but keeping the Hilbert space factorized), we conclude that to a high degree of accuracy the correlation energy obtained in the APSG scheme can be calculated as the sum of independent pair contributions [44]. This result also justifies our using the same symbol for the coefficients occurring in Ψ_{APSG} and Ψ_R. As the variational equation from which they are obtained is (practically) the same, they also should be (practically) the same. This means that the correction functions $u_{R\bar R}$ can be calculated independently and that the ε_R in the sense of Section III.A are identical with the E_{corr}^R.

We now examine the wave function $\Psi_{(2)}$ obtained from $\Psi_{\text{corr}}^{\text{APSG}}$ on truncating it to double substitutions only (and keeping the Hilbert space factorized). Since we do not know whether the coefficients will be the same, we use a different letter for them

$$\Psi_{(2)} = N'\{\Phi + \sum_{R,k} b_k^R\, \Phi_{R\bar R}^{k\bar k}\} \tag{84}$$

Here N' means a normalization factor which turns out to be

$$N' = [1 + \sum_{\substack{R,k \\ (>1)}} (b_k^R)^2]^{-\frac{1}{2}} \tag{85}$$

The correlation energy accounted for with $\Psi_{(2)}$

$$E_{\text{corr}}^{(2)} = E_{(2)} - E_0 \tag{86}$$

can be written in the form

$$E_{\text{corr}}^{(2)} = \sum_{R,k} g_k^R E_k^R + \sum_R {\sum_{k,l}}' f_k^R f_l^R (R_k R_l | R_l R_k) \tag{87}$$

with

$$f_k^R = \begin{cases} N' b_k^R & \text{for } k > 1 \\ \\ N' & \text{for } k = 1 \end{cases} \tag{88}$$

$$g_k^R = \begin{cases} (f_k^R)^2 & \text{for } k > 1 \\ \\ (f_k)^2 - 1 & \text{for } k = 1 \end{cases} \tag{89}$$

$$E_k^R = 2 H_{kk}^R + (R_k R_k | R_k R_k) \tag{90}$$

On the other hand $\sum_R E_{\text{corr}}^R$ which is practically identical with E_{APSG} can be written as

$$E_{\text{APSG}} \approx \sum_R E_{\text{corr}}^R = \sum_{R,k} a_k^R E_k^R + \sum_R {\sum_{k,l}}' d_k^R d_l^R (R_k R_l | R_l R_k) \tag{91}$$

Formally (87) and (91) are exactly the same. The only important difference is the different normalization of the d_k^R and the f_k^R, in fact

$$\sum_{k=1} (d_k^R)^2 = 1$$

$$(N_R)^2 = \sum_{k=1}^{\infty} (f_k^R)^2 = (N')^2 \left[1 + \sum_{k>1} (b_k^R)^2\right] = \left[1 + \sum_R \sum_{k>1} (b_k^R)^2\right]^{-1} \times$$

$$\times \left[1 + \sum_{k>1} (b_k^R)^2\right] \tag{92}$$

If we assume proportionality between the f_k^R and d_k^R and see their difference only in the different normalization, we find that

$$E_{\text{corr}}^{(2)} = \sum_R N_R^2 \cdot E_{\text{corr}}^R \tag{93}$$

The correlation energy obtained with $\Psi_{(2)}$ is not equal to the sum of the individual pair contributions E_{corr}^R, but to a sum of contributions $N_R \cdot E_{\text{corr}}^R$ with a factor $N_R < 1$, which depends on the normalization of the other pairs. If the coefficients are small enough (and the number of pairs is small), one can expand N_R and get.

$$N_R = [1 + \sum_{k>1} (b_k^R)^2]^{\frac{1}{2}} \ [1 + \sum_S \sum_{k>1} (b_k^S)^2]^{-\frac{1}{2}}$$

$$\approx [1 - \sum_{\substack{S \\ (\neq R)}} \sum_{k>1} (b_k^R)^2]^{\frac{1}{2}} \approx 1 - \frac{1}{2} \sum_{\substack{S \\ (\neq R)}} \sum_{k>1} (b_k^R)^2 \qquad (94)$$

In the case of the Be ground state [44] with configuration $K^2 L^2$, one has $N_K^2 \approx 0.95$, $N_L^2 \approx 0.99$ and therefore since $E_{corr}^L \approx F_{corr}^K$, $E_{corr}^{(2)} \approx 0.97 \ E_{corr}^{APSG}$.

Ψ_{APSG} differs from $\Psi_{(2)}$ in that it includes the unlinked clusters. We therefore have the interesting result that inclusion of the unlinked clusters allows us to treat the pairs independently. The sum of the independently calculated pair correlation energies is an approximation to the correlation energy of the APSG wave function but not to that of a wave function limited to double substitutions.

One must, however, note that this result only holds if the Hilbert space is factorized (strong orthogonality of the geminals) and that the difference between $E_{corr}^{(2)}$ and E_{corr}^{APSG} is only pronounced if there are coefficients $d_k^R (k > 1)$ which are relatively large (like the coefficient ~ -0.3 of the $1s^2 2p^2$-configuration in the Be ground state, due to near degeneracy). If the coefficients are sufficiently small (say $|d_k^R| < 0.1$) and if the number of the electron pairs is small enough, the difference between E_{corr}^{APSG} and $E_{corr}^{(2)}$ will not exceed a few per cent.

The most important statement concerning the unlinked clusters is that their omission leads to an increasing error for an increasing number of electrons In the limit $n \to \infty$ N_R and with it $E_{corr}^{(2)}$ goes to zero and $\Psi_{(2)}$ becomes useless. If one claculates both a molecule M and its dimer M_2 with $\Psi_{(2)}$ *i.e.* without unlinked clusters one will not get a reliable dimerization energy. E_{corr}^{APSG} on the other hand has the correct dependence on n even in the limit $n \to \infty$.

E. The Independent Electron-Pair Approximation (IEPA) for the Intra-Pair Correlation

As we have stated, the APSG scheme has two disadvantages, one of them being that the equations for the different pairs are coupled, the other that it accounts for only a (possibly small) part of the correlation energy.

We therefore need both to generalize and to simplify the scheme. One simplification has already been justified in the last section, namely that we can calculate the correlation contribution of the different pairs independently (but keeping the Hilbert space factorized). A very promising generalization seems to be to relax the factorization of the Hilbert space (in other words, the strong orthogonality between the geminals). We do not start from the energy expectation value of a general APG wave function, $i.e.$ from the counterpart of APSG without strong orthogonality, but we examine first how the results at the end of the last section are modified if we relax the factorization of the Hilbert space.

First, we see easily that $E_{corr}^{(2)}$ of (87) remains formally the same. The reason is that the Slater determinants in $\Psi_{(2)}$ of (84) are mutually orthogonal even if (64) does not hold. For their orthogonality, it is sufficient that the strongly occupied ϕ_1^R are orthogonal to each other.

We can therefore vary the pair correlation functions for the different pairs completely freely (without factorization of the Hilbert space) and calculate the correlation energy corresponding to $\Psi_{(2)}$ by the use of (93). This increased flexibility in the variation will, of course, lower the energy relative to a $\Psi_{(2)}$ with factorization of the Hilbert space.

We now ask whether inclusion of the unlinked clusters has the same effect as in the last section, namely to allow a direct addition of the pair correlation energies, calculated independently. The answer is not so easy as in the case where the Hilbert space is factorized.

We have in fact to compare the energy of an APG function (Antisymmetrized Product of Geminals, without strong orthogonality) with the sum of individual pair energies (without factorization of the Hilbert space). The difference

$$\Delta E = E_{corr}^{APG} - \sum \tilde{E}_{corr}^R \tag{95}$$

may be called the "additive error". We use here a tilde on $\tilde{E}_{corr}^R$ to indicate that (in contrast to the last section) no factorization of the Hilbert space is imposed for the independent calculation of any pair correlation energy.

The full expression for the additivity error ΔE is much more lengthly than for its counterpart in the case of a factorized Hilbert space given by (81). We are therefore not writing down an explicit expression for ΔE, we are just indicating which kind of terms give the most important contributions to E [44].

One can classify the contributions to ΔE according to the order in which they depend on the coefficients d_k^R ($k > 1$) of weakly occupied natural orbitals of the R-th pair. Most contributions are of fourth (or higher) order in the d_k^R and also of second (or higher) order in the overlap integrals S_{ij}^{RS} of weakly occupied natural orbitals of different pairs, and are therefore negligible.

There are, however, also contributions that are of the third order in the d_k^R and of the first order in S_{ij}^{RS}, namely terms like

$$- d_1^R \sum_{k,j\,(>1)} (d_j^S)^2\, d_k^R\, S_{kj}^{RS}\, (R_1 R_k | R_1 S_j) \tag{96}$$

In the case of the Be ground state, a term of this kind enters with a factor 4. In this case we have a relatively large coefficient of a weakly occupied orbital, namely

$$d(2 \text{ pL}) \approx - 0.3$$

whereas the largest coefficient of a weakly occupied orbital in the K shell is[c]

$$d(2 \text{ pK}) \approx - 0.03$$

The overlap integral between 2 pK and 2 pL is

$$S(2 \text{ pK}, 2 \text{ pL}) \sim 0.3$$

and the two-electron integral to be combined with these values

$$(1 \text{ sK}, 2 \text{ pK}|1 \text{ sK}, 2 \text{ pL}) \approx 0.4 \text{ a.u.}$$

This leads to

$$- 4 \, d(1 \text{ sK}) \, [d(2 \text{ pL})]^2 \, d(2 \text{ pK}) \, S(2 \text{ pL}, 2 \text{ pK}) \, (1 \text{ sK}, 2 \text{ pK}|1 \text{ sK}, 2 \text{ pL})$$
$$\sim 1{,}6.10^{-3} \text{ a.u.} \tag{97}$$

which[c] is about 1% of the correlation energy of the Be ground state. Since the other contributions to ΔE are negligible with respect to this one, we can conclude that the additivity error for the Be ground state is roughly 1% of the correlation energy.

If one adds up the pair-correlation energies calculated for the K and L shells separately, one gets about 1% more correlation energy than if one had correctly calculated the expectation value of an APG function.

This example gives us some idea about the conditions under which the additivity error may be negligibly small. Two quantities seem to be crucial.

1. The coefficients d_k^R of the weakly occupied natural orbitals of the pairs. The smaller the d_k^R, the smaller the additivity correction.

If *e.g.* $d(2 \text{ pL})$ were ≈ -0.1 rather than -0.3, the expression (97) would be reduced by a factor of ~ 10.

---

[c] In this example d is the coefficient of a normalized p^2 configuration, which is actually the sum of $(p_x)^2$, $(p_y)^2$ and $(p_z)^2$. If this is explicitly taken into account, the result in (97) must be divided by a factor $\sqrt{3}$.

2. The overlap integrals $S_{kl}^{RS} = S(Rk, Sl)$ between weakly occupied orbitals of different pairs. If the Hilbert space were factorized, all these S_{kl}^{RS} would vanish and the additivity error would be negligible.

One may ask whether these two kinds of crucial quantities are fixed for a given state or whether there is some freedom in their choice. In fact there is a possibility of making the overlap integrals S_{kl}^{RS} as small as possible if one expresses the leading Slater determinant Φ in terms of localized rather than canonical orbitals. If ϕ_1^R and ϕ_1^S are localized in different regions of space, ϕ_1^R and ϕ_1^S will also be localized in different regions of space, namely in the same region as their strongly occupied orbitals, and the S_{kl}^{RS} will be very small.

One way of reducing the additivity error is therefore to use localized orbitals.

For systems with well localized orbitals and with small coefficients of the weakly occupied orbitals, $e.g.$ methane [45,46], the additivity errors are supposed to be smaller than about 1% of the correlation energy, at least, as long as one is concerned with the intra-pair correlation only.

It has been shown [47] that the additivity errors may become enormous if one goes from a localized to a delocalized description.

F. The Independent Electron-Pair Approximation for Intra and Interpair Correlation

The next step in a generalization of the independent electron-pair correlation is to include the interpair correlation contributions as well. The recipe of the general independent electron-pair approximation (IEPA) is hence to calculate independently the correlation-energy contributions E_{corr}^{ij} for any pair of spin orbitals, $i.e.$ the correlation energy accounted for by the wave function

$$\Psi_{ij} = c_0^{ji} \phi + \sum_{a<b} c_{ij}^{ab} \phi_{ij}^{ab} \tag{98}$$

and to regard the sum

$$\sum_{i<j} E_{corr}^{ij} = E_{corr}^{IEPA} \tag{99}$$

as an approximation to the true correlation energy. The contribution E_{corr}^{ij} may be classified as intrapair correlation energy if ϱ_i and ϱ_j have the same spatial but different spin factors and otherwise as interpair.

This recipe was suggested some time ago by Sinanoglu [40] and Nesbet [34]. It differs from the one outlined in Section III.E in that we now consider $\left(\frac{n}{2}\right) = \frac{n(n-1)}{2}$ rather than $\frac{n}{2}$ pairs of spin orbitals.

There are two main reasons why this general independent electron-pair approximation (IEPA) has not been too popular for some time.

1. It is much harder to calculate or estimate the additivity error for the general IEPA than for the IEPA limited to the intrapair correlation. In some cases, *e.g.* for the neon ground state, this error can be of the order of 10% of the total correlation energy (*i.e.* E_{corr}^{IEPA} roughly 110% of the true correlation energy). After this discovery [48,49], certain cases[50] where 100% of the experimental correlation energy was obtained were attributed to compensating errors. The situation is becoming better understood.

2. There was still the hope that, at least for molecules with localized bonds, intrapair correlation energy might be the main contribution to the total correlation energy.

This hope turned out to be vain and it became clear [40,45,46] that if molecular calculations are to account for electron correlation, they have to allow for the interpair correlation as well. So a systematic method to calculate the intra- and interpair correlation contribution has been programmed and has since been applied by the author's research group to several small molecules. The first study of this kind was published by Jungen and Ahlrichs [46]. Analogous calculations for atoms (though not based on the direct calculation of approximate natural orbitals) had been performed before, mainly by Nesbet [50].

As long as the problem of the additivity errors is not completely solved one can at least regard the IEPA as a method that allows one with very little effort to improve the accuracy of quantum mechanical calculations by at least one order of magnitude relative to the Hartree-Fock scheme. This is a substantial advantage if the variation principle is not strictly obeyed.

G. The PNO-CI and CEPA Methods[d]

The IEPA-method has the advantage of being formally rather simple and not very time-consuming in computation, but it has two main drawbacks.

1. It does not furnish an upper bound to the energy, it is not a genuine variational method

2. It does not give information about the 'additivity error' *i.e.* the difference between the IEPA-energy and the exact energy to which IEPA is an approximation.

---

[d] This Section was added in June 1973.

At first glance one may think that the two drawbacks are one and the same. In *practice* it is however not possible to improve the method such as to eliminate the two drawbacks simultaneously. On one hand one can, without great difficulty, using the pair correction functions $\tilde{u}_{ij}$ obtained by the IEPA scheme, compute an upper bound to the true energy, namely $E_{(2)}$ of Eq. (62). This $E_{(2)}$ is however, in some respect poorer than $\tilde{E}$ of Eq. (63) calculated with the same $\tilde{u}_{ij}$ i.e. the IEPA energy. In particular as was mentioned in Section III. D, $E_{(2)}$ (unlike $\tilde{E}_{\mathrm{IEPA}}$) has the incorrect dependence on the number n of electrons and becomes increasingly poorer the larger n is. In order to get an energy that has the correct n-dependence one has to take care of the 'unlinked clusters'. Unfortunately this is much less straightforward in the general case where one considers all pairs than it is in the special case treated in Section III. D and E. One can formulate the respective wave function most elegantly in the form [38]

$$\Psi = \exp \left(\sum_{i<j} \hat{S}_{ij} \right) \Phi \tag{100}$$

The operator $\hat{S}_{ij}$ is defined such that its application to the Slater determinant Φ yields

$$\hat{S}_{ij} \Phi = \sum_{a<b} c_{ij}^{ab} \, \Phi_{ij}^{ab} \tag{101}$$

Any $\hat{S}_{ij}$ is characterized by the coefficients c_{ij}^{ab} and contains the same information as the u_{ij}. The exponential function in (100) is to be understood as the corresponding power series and some other precautions have to be taken. We are not going into details here since so far no one has succeeded in deriving a rigorous and practicable *variational* treatment based on the ansatz (100). A computational scheme for molecules based on (100) and using diagram techniques has been proposed by Cizek *et al.*[67] (called CMET for Coupled Many-Electron-Theory) but it is formally very complicated and has not yet been applied in the context of large-scale ab initio calculations. CMET is not a variational method and does not give a rigorous upper bound, but in the cases studied so far (where an exact reference calculation was possible) it led to results that were rather close to the exact values.

Recently W. Meyer [49a] has proposed two rather straightforward extensions of the IEPA method that proved to be very useful in practical calculations. The first one termed PNO—CI (Pseudo-Natural-Orbital Configuration Interaction) eliminates the first drawback of IEPA mentioned in the beginning of this section, the second one, called CEPA (for Coupled-Electron-Pair Approximation) improves IEPA with respect to the second drawback. The relation between the IEPA, PNO—CI and CEPA methods is most easily explained in the following way.

Take the 'leading' Slater determinant Φ_0 and the doubly substituted ones Φ_{ij}^{ab} as basis functions for the expansion of wave function

$$\Psi_{\rm CI} = \Phi_0 + \sum_{i<j} \sum_{a<b} c_{ij}^{ab} \; \Phi_{ij}^{ab} \tag{102}$$

We normalize $\Psi_{\rm CI}$ such that the coefficient of Φ_0 is equal to 1. The spin orbitals φ_a, φ_b into which one 'excites' are taken as the natural spin orbitals of the respective IEPA pair correction function $\tilde{u}_{ij}$ [22]. This will be the case for all of the methods considered and we shall not stress this again. One ought to add the abbreviation PNO (for Pair-Natural-Orbitals, sometimes interpreted as Pseudo-Natural-Orbitals) to the terms IEPA and CEPA as well, $i.e.$ to speak of IEPA-PNO and CEPA-PNO rather than just of IEPA and CEPA. Using the PNO's one gets mutually orthogonal Φ_{ij}^{ab} though the PNO's of different pairs are nonorthogonal. Also the explicit expressions for $H_{\rm AB}$ are not too complicated.

The best coefficients c_{ij}^{ab} in the CI-wave function (102) — that contains only double substitutions and no unlinked clusters — are obtained as components of the eigenvektor $\vec{C}_{\rm CI}$ that belongs to the lowest eigenvalue $E_{\rm CI}$ of the Hamiltonian matrix H in the representation of Φ_0 and the Φ_{ij}^{ab}. To simplify the argument we now replace the double index ij by a single index, that counts pairs, we further assume that there are only two different pairs, A and B. The Matrix H has than a block structure of the form

$$H = \begin{pmatrix} E_0 & H_{\rm OA} & H_{\rm OB} \\ H_{\rm AO} & H_{\rm AA} & H_{\rm AB} \\ H_{\rm BO} & H_{\rm BA} & H_{\rm BB} \end{pmatrix} \tag{103}$$

where E_0 is a single number (the energy of Φ_0), $H_{\rm AB}$ has $e.g.$ the dimension $n_{\rm A} \times n_{\rm B}$ if there are $n_{\rm A}$ Slater determinants obtained from replacing the pair A in Φ_0 by two PNO's of $\tilde{u}_{\rm A}$ and $n_{\rm B}$ defined analogously.

The eigenvector of H has a block structure as well

$$\vec{C}_{\rm CI} = (1, \vec{C}_{\rm A}^{\rm CI}, \vec{C}_{\rm B}^{\rm CI}) \tag{104}$$

and it holds, of course, that the PNO-CI energy is given by

$$E_{\rm CI} = \; <\vec{C}_{\rm CI}|H|\vec{C}_{\rm CI}> / <\vec{C}_{\rm CI}|\vec{C}_{\rm CI}> \tag{105}$$

W. Kutzelnigg

The IEPA-method consists in calculating the eigenvalues $E_0 + \varepsilon_A^{\text{IEPA}}$ and $E_0 + \varepsilon_B^{\text{IEPA}}$ and the eigenvektors $(1, \vec{C}_A^{\text{IEPA}})$ and $(1, \vec{C}_B^{\text{IEPA}})$ of

$$\mathbf{H_A} = \begin{pmatrix} E_0 & H_{\text{OA}} \\ H_{\text{AO}} & H_{\text{AA}} \end{pmatrix} \text{ and } \mathbf{H_B} = \begin{pmatrix} E_0 & H_{\text{OB}} \\ H_{\text{BO}} & H_{\text{BB}} \end{pmatrix} \tag{106}$$

rather than those of $\mathbf{H}$.

The energy $E_{(2)}$ of Eq. (62) which (like E_{CI}) is an upper bound to the energy, is in this case

$$E_{(2)} = \; <\vec{C}_{\text{IEPA}} |\mathbf{H}| \vec{C}_{\text{IEPA}} > / <\vec{C}_{\text{IEPA}} |\vec{C}_{\text{IEPA}} > \tag{107}$$

with

$$\vec{C}_{\text{IEPA}} = (1, \vec{C}_A^{\text{IEPA}}, \vec{C}_B^{\text{IEPA}}) \tag{108}$$

Obviously E_{CI} is of the same form as $E_{(2)}$, in fact E_{CI} is the lowest of all expectations values of this form. We expect that in 'normal' cases *i.e.* if $\mathbf{H}_{\text{AB}}$ is not too large $E_{(2)}$ and E_{IEPA} do not differ very much.

Using the fact that the IEPA-coefficients make the expectation values of $\mathbf{H_A}$ and $\mathbf{H_B}$ stationary separately, we can express $E_{(2)}$ (which 'normally' differs little from E_{CI}) in terms of $\varepsilon_A^{\text{IEPA}}$ and $\varepsilon_B^{\text{IEPA}}$

$$E_{(2)} = E_0 + \varepsilon_A^{\text{IEPA}} \cdot \frac{1 + C_A^2}{1 + C_A^2 + C_B^2} + \varepsilon_B^{\text{IEPA}} \cdot \frac{1 + C_B^2}{1 + C_A^2 + C_B^2}$$
$$+ \frac{2 <\vec{C}_A |H_{\text{AB}}| \vec{C}_B >}{1 + C_A^2 + C_B^2} \tag{109}$$

If the 'off-diagonal' block vanishes we get essentially Eq. (93) Considering the discussion of Section III. D and E we expect that inclusion of the unlinked clusters should have the main effect to remove the 'normalization factors' and to lead to something like

$$E_{\text{LC}} = E_0 + \varepsilon_A^{\text{IEPA}} + \varepsilon_B^{\text{IEPA}} + \frac{2 <\vec{C}_A |H_{\text{AB}}| \vec{C}_B >}{(1 + C_A^2)(1 + C_B^2)} \tag{110}$$

This expression will have the correct n-dependence at least for vanishing $\mathbf{H}_{\text{AB}}$ though we cannot expect that it is rigorous. It will 'normally' be more physical than $E_{(2)}$, but not be an upper bound to the energy.

In the same way as we came from $E_{(2)}$ to E_{CI} we can go from E_{LC} to E_{CEPA} in regarding E_{LC} as an expectation value (namely as an approximation to the expectation value of the function (100)) and making it stationary with respect to variation of the $\vec{C}_A$ and $\vec{C}_B$. This leads to two coupled inhomogeneous eigenvalue equations.

$$(\boldsymbol{H}_A - E_0 - \varepsilon_A^{\text{CEPA}}) \begin{pmatrix} 1 \\ \vec{C}_A \end{pmatrix} = \begin{pmatrix} O \\ \boldsymbol{H}_{AB} \ \vec{C}_B \end{pmatrix}$$

$$(\boldsymbol{H}_B - E_0 - \varepsilon_B^{\text{CEPA}}) \begin{pmatrix} 1 \\ \vec{C}_B \end{pmatrix} = \begin{pmatrix} O \\ \boldsymbol{H}_{BA} \ \vec{C}_A \end{pmatrix} \tag{111}$$

Note that the new eigenvalues $\varepsilon_A^{\text{CEPA}}$ are 'additive' in the sense that

$$E_{\text{CEPA}} = E_0 + \varepsilon_A^{\text{CEPA}} + \varepsilon_B^{\text{CEPA}} \tag{112}$$

So they are 'good' pair energies in the sense of Section III. B.

The derivation of CEPA given here [68] is not identical with but equivalent to the one given by Meyer [49a]. It is not very relevant in this context that Meyer has proposed two different version of CEPA.

Whereas the IEPA-scheme means physically that one treats one electron pair in the Hartree-Fock field of the other electrons, CEPA means that each pair is treated in the fields of the correlated other electrons which is somewhat more physical. Through the explicit consideration of the $\boldsymbol{H}_{AB}$ block the interaction between the different pair correlations is accounted for. Though a rigorous justification of the CEPA-method is still lacking and probably not possible the practical applications [49a] have so far been very satisfactory. One main advantage of CEPA as compared with IEPA is that it can also be used with delocalized orbitals and the results are nearly invariant with respect to a unitary transformation of the occupied orbitals.

H. Other Methods Based on the Idea of Electron Pairs

For the sake of completeness we can mention some other computational schemes, different from those explained here, that are also based on the idea of electron-pair correlation. There is *e.g.* the use of the following ansatz for the wave function

$$\Psi(1,2 \ldots n) = A\{\omega(1,2)\ \omega(3,4) \ldots \omega(n-1,n)\} \tag{113}$$

i.e. a special case of an APG function in which all geminals are identical. This ansatz has been called APIG (*a*ntisymmetrized *p*roduct of *i*dentical *g*eminals) or AGP (*a*ntisymmetrized *g*eminal *p*ower) or n-projected BCS function (because of its relation to the Bardeen-Cooper-Schrieffer theory of superconductivity). This kind of wave function has been introduced by

Nakamura [52] in the context of the theory of superconductivity and has played an interesting role in relation to the n-representability problem [53]. Finally this ansatz has been applied to molecular calculations by Bratoz and Durand [54]. In spite of some encouraging results, the use of (100) for the treatment of electron correlation in molecules seems to be a dead end.

We have then to mention perturbation theoretical methods. The full Hamiltonian is split into aseparable Hamiltonian H_0 and a perturbation V.

$$H = H_0 + V$$

where H_0 may be the bare nuclear Hamiltonian or the Hartree-Fock Hamiltonian. The unperturbed solution (*i.e.* the eigenfunction of H_0) is a single Slater determinant Φ and the perturbation correction to both E and Φ can be formally expanded in terms of powers of V. The first-order correction E_1 to the energy vanishes if Φ is the Hartree-Fock wave function and the second-order correction is exactly the sum of pair contributions. So in a second-order perturbation treatment pair energies come out automatically [55]. This is no longer the case if one goes over to higher orders in perturbation theory.

An alternative use of perturbation theory together with the idea of electron pairs is the method of Malrieu, Diner and Clavery abbreviated as PCILO [56] (*p*erturbative *c*onfiguration *i*nteraction based on *l*ocalized *o*rbitals). One uses a zeroth-order wave function, one in which each electron pair is in a fully localized LCAO-MO. All delocalization and correlation corrections are then treated by perturbation theory (not limited to second order).

There are finally attempts to apply diagrammatic techniques of manybody perturbation theory [57], with the summation of certain diagrams to infinite order, to the correlation problem in atoms and molecules. A close relationship between this kind of approach and the independent electronpair approximation has been demonstrated [58].

IV. Outline of Some Results

Electron correlation energies for small molecules have been calculated either by the independent electron-pair approximation (IEPA) or by configuration interaction (CI). Brute force CI in general did not give too good results; CI calculations gave a substantial part of the correlation energy only where the weakly occupied orbitals had been optimized somehow. Here one must mention the calculations on diatomic molecules by Bender and Davidson [59] and by Wahl *et. al.* [60].

For molecules with more than two atoms, most calculations that account for the main part of electron correlation are based on IEPA. PNO—CI- and

CEPA calculations on the H_2O and CH_4 molecules have recently been performed by W. Meyer [61,49a].

We are not going to discuss the results of individual calculations, nor shell we explain the computational procedure in detail [62]. We prefer to outline some results of more general interest.

Let us first look at the values of intra- and interpair correlation of the valence-shell energies for the molecules LiH, BeH_2, BH_3, CH_4, BH_2^+ and CH_3^+

	$-\varepsilon_R$	n	$-n \cdot \varepsilon_R$	$-\varepsilon_{RS}$	$\dfrac{n(n-1)}{2}$	$\dfrac{-n(n-1)\varepsilon_{RS}}{2}$
LiH	0.0340	1	0.0340		0	0.0
BeH_2	0.0334	2	0.0668	0.0054	1	0.0054
BH_3	0.0327	3	0.0971	0.0096	3	0.0288
CH_4	0.0300	4	0.1200	0.0154	6	0.0924
BH_2^+	0.0340	2	0.0680	0.0113	1	0.0113
CH_3^+	0.0313	3	0.0939	0.0167	3	0.0501

In this table any calculated ε is about 90% of the exact independent-pair contributions, and n is the number of valence pairs (*i. e.* of bonds).

We note that the intrapair correlation energy ε_R is roughly 0.032 a.u. for a XH bond. The precise numerical value depends on two items. First ε_R depends on the XH distance. ε_R does not change much between $R_{XH} = 0$ and the equilibrium distances of these molecules; for larger distances $|\varepsilon_R|$ increases considerably [63]. This dependence on distance is not relevant for the change on going from LiH to CH_4. For this change the following mechanism is essential: the weakly occupied natural orbital must be strongly orthogonal to the strongly occupied ones. So the more strongly occupied orbitals there are, the less is the flexibility for weakly occupied ones. In LiH all the 2p-AC's of the heavy atoms are "available" for correlation, in BeH_2 only two of them, in BH_3 one, and in CH_4 none is "available". The importance of this exclusion effect was probably first postulated by Sinanoglu [40].

An individual interpair (or interbond) correlation energy ε_{RS} increases considerably in absolute value on going from LiH to CH_4. The reason is that the bond angle decreases steadily from LiH to CH_4 and so the electrons in the different bonds come increasingly closer to each other. The closer two pairs are, the stronger is their interpair correlation interaction. Since the number of interpair terms increase (compared to the number of pairs) in the same series, one finds a substantial increase of the interpair contribution to the total valence-shell correlation energy on going from LiH (0%) to CH_4 (almost 50%).

The interpair contributions are larger in BH_2^+ and CH_3^+ than in their respective isoelectronic molecules BeH_2 and BH_3. One can explain this by noting that, due to the higher nuclear charge, the electrons in the bonds come closer to the nuclei and hence closer to each other.

These calculations did not include the intrapair correlation energy of the K-shell electrons and the interpair correlation between the K shell and the valence shell. Calculations on some test systems [63] have shown that these contributions do not change appreciably on going from the isolated atoms to the molecule and hence have little influence on binding energies.

One noteworthy result is related to the transferability of localized electron pairs [58,59]. One sees by looking at maps of the electron density that, e.g. the BeH bonds in BeH, BeH_2 and Be_2H_4 (terminal bond), are very much the same. Now it turns out that the intrapair correlation energies calculated for such a BeH bond in any of these systems is practically the same. Correlation energy contributions are transferable for the same bond in different molecules. This result is, of course, rather important for the understanding of the binding energies of, say, saturated hydrocarbons.

One also finds that interpair correlation energies are transferable if the two bonds and their relative position (e.g. angle) are the same. For large distances between two electron pairs the interpair correlation energies follow asymptotically the London formula for dispersion energies, i.e. they go like R^{-6} [66]. In fact the same formalism can be used to calculate interpair correlation energies within a molecule and dispersion energies between different molecules. This dispersion interaction is a correlation effect of the interpair type.

With the method based on the independent-pair approximation (IEPA) and the direct calculation of approximate natural orbitals of the electrons that has allowed us to obtain the results just discussed it is also possible to find an answer to the problem of the dimerization energy of BH_3 [7]. The experimental values are somewhat uncertain, since they range between 20 and 60 kcal/mol. Earlier Hartree-Fock calculations never gave more than 10 kcal/mol. Gelus et al. [7] have confirmed these Hartree-Fock results, but also estimated that the correlation energy changes by about 25 kcal/mol on dimerization, so that the probable value of 35 kcal/mol for the dimerization energy is mainly due to the change in correlation. A detailed analysis showed that the intrapair correlation energy does not change appreciably. The interpair correlation energy changes greatly for the simple reason that in B_2H_6 the number of pairs of neighbour bonds is eleven but only six in 2 BH_3 (3 in BH_3). We conclude that for the correlation energy on molecule (or bond) formation to be approximately constant, it is necessary not only that the number of coupled electron pairs remains the same, but also the number of neighbour pairs of pairs. Recent SCF calculations with a more extended basis set [69,68] lead to a SCF-binding energy of roughly 20 kcal/mol. There is also recent evidence that the change in correlation energy is somewhat smaller than estimated in Ref.[7]. Another case where similar correlation contribution as in B_2H_6 play a crucial role is that of the classical vs. the nonclassical structures of $C_2H_3^+$ and $C_2H_5^{+[70]}$.

Acknowledgement. The work reported here was sponsored by the Deutsche Forschungsgemeinschaft and the Fonds der Chemischen Industrie. Some of the material was presented at the summer school on Theoretical Chemistry in Linz/Austria 1971. The author thanks particularly Dr. R. Ahlrichs and Dr. V. Staemmler for their contribution to the work summarized here and to Dr. W. Meyer for discussions.

References

1) Wigner, E. P., Seitz, F.: Phys. Rev. *43*, 804 (1933); *46*, 509 (1934).

2) Löwdin, P. O.: Advan. Chem. Phys. *2*, 207, (1959).

3) Nesbet, R. K., Barr, T. L., Davidson, E. R.: Chem. Phys. Letters *4*, 203 (1969).

4) Veillard, A.: Chem. Phys. Letters *3*, 128, 565 (1969).

5) Diercksen, G. H. F.: Theoret. Chim. Acta *21*, 335 (1971).

6) Clementi, E.: J. Chem. Phys. *46*, 3851 (1967).

7) Gelus, M., Ahlrichs, R., Staemmler, V., Kutzelnigg, W.: Chem. Phys. Letters *7*, 503 (1970).

8) Møller, C., Plesset, M. S.: Phys. Rev. *46*, 618 (1934).

9) Grimaldi, F., Lecourt, A., Moser, C.: Intern. J. Quantum Chem. *1 S*, 153 (1970). Stevens, R. M., Karplus, M.: J. Chem. Phys. *49*, 1094 (1968).

9a) v. Herigonte, P.: Struct. Bonding, *12*, 1 (1972)

10) Ruedenberg, K.: Rev. Mod. Phys. *34*, 326 (1962).

11) McWeeny, R.: Proc. Roy. Soc. (London) *A 253*, 242 (1959); Rev. Mod. Phys. *32*, 335 (1960).

12) Kutzelnigg, W.: Z. Naturforsch. *18 a*, 1058 (1963); 168 (1965).

13) McWeeny, R., Kutzelnigg, W.: Intern. J. Quantum Chem. *2*, 187 (1968).

14) Kutzelnigg, W., Del Re, G., Berthier, G.: Phys. Rev. *172*, 49 (1968).

15) Kato, T.: Commun. Pure Appl. Math. *10*, 151 (1957).

16) Bingel, W. A.: Theoret. Chim. Acta *5*, 341 (1966).

17) Hurley, A. C., Lennard-Jones, J., Pople, J. A.: Proc. Roy. Soc. (London) *A 220*, 446 (1953).

18) Löwdin, P. O., Shull, H.: Phys. Rev. *101*, 1730 (1956).

19) Schwartz, C.: Phys. Rev. *126*, 1015 (1962).

20) Lakin, W.: J. Chem. Phys. *43*, 2954 (1965).

21) Ahlrichs, R., Kutzelnigg, W., Bingel, W. A.: Theoret. Chim. Acta *5*, 289 (1966).

22) Kutzelnigg, W.: Theoret. Chim. Acta *1*, 327 (1963).

23) Bopp, F.: Z. Phys. *156*, 348 (1959).

24) Yang, C. N.: Rev. Mod. Phys. *34*, 694 (1962).

25) Sasaki, F.: Phys. Rev. *138 B*, 1338 (1965).

26) Coleman, A. J.: Rev. Mod. Phys. *35*, 668 (1963).

27) Watanabe, S.: Z. Phys. *113*, 482 (1930).

28) Mayer, J. E.: Phys. Rev. *100*, 1579 (1955).

29) Garrod, C., Percus, J. K.: J. Math. Phys. *5*, 1756 (1964).

30) Davidson, E. R.: J. Math. Phys. *10*, 725 (1969) — Mc Rae, W. B., Davidson, E. R.: J. Math. Phys. *13*, 1527 (1972).

31) Peat, F. D., Brown, R. J. C.: Can. J. Phys. *44*, 1344 (1969). — Peat, F. D.: Can. J. Phys. *45*, 847 (1967); Phys. Rev. *173*, 67 (1968). — Bender, C. F., Davidson, E. R., Peat, F. D.: Phys. Rev. *174*, 75 (1968).
32) Garrod, C., Rosina, M.: J. Math. Phys. *10*, 1855 (1969).
33) Kutzelnigg, W.: Paper given at the workshop on density matrices in Kington/Ontario, Sept. 1970.
34) Nesbet, R. K.: Phys. Rev. *109*, 1632 (1958); Advan. Chem. Phys. *9*, 321 (1965).
35) Kutzelnigg, W., Smith, V. H., Jr.: Chem. Phys. *41*, 896 (1964).
36) Primas, H.: In: Modern quantum chemistry. (O. Sinanoglu, ed.). New York: Academic Press 1965.
37) Brenig, W.: Nucl. Phys. *4*, 363 (1957).
38) Coester, F.: Nucl. Phys. *7*, 421 (1958). — Coester, F., Kümmel, H.: Nucl. Phys. *17*, 477 (1960). — Kümmel, H.: Nucl. Phys. *22*, 177 (1961).
39) Brout, R.: Phys. Rev. *111*, 1324 (1958).
40) Sinanoglu, O.: J. Chem. Phys. *36*, 706, 3198 (1962).
41) Kutzelnigg, W.: J. Chem. Phys. *40*, 3640 (1964).
42) Miller, K. J., Ruedenberg, K.: J. Chem. Phys. *48*, 3418 (1968).
43) Mehler, E. L., Ruedenberg, K., Silver, D. M.: J. Chem. Phys. *52*, 1181 (1970).
44) Ahlrichs, R., Kutzelnigg, W.: J. Chem. Phys. *48*, 1819 (1968).
45) Ahlrichs, R., Kutzelnigg, W.: Chem. Phys. Letters, *1*, 651 (1968).
46) Jungen, M., Ahlrichs, R.: Theoret. Chim. Acta *17*, 339 (1970).
47) Davidson, E. R., Bender, C. F.: J. Chem. Phys. *49*, 465 (1968). — Bender, C. F., Davidson, E. R.: Chem. Phys. Letters *3*, 33 (1969).
48) Barr, T. L., Davidson, E. R.: Phys. Rev. *A 1*, 644 (1970).
49) Viers, J. W., Harris, F. E., Schaefer III, H. F.: Phys. Rev. *A 1*, 24 (1970).
49a) Meyer, W.: J. Chem. Phys. *58*, 1017 (1973).
50) Nesbet, R. K.: Phys. Rev. *155*, 51, 56 (1967); *175*, 2 (1968).
51) Kutzelnigg, W.: to be published.
52) Nakamura, K.: Progr. Theoret. Phys. *21*, 713 (1959). — Blatt, J. M.: Progr. Theoret. Phys. *23*, 447 (1960).
53) Coleman, A. J.: J. Math. Phys. *6*, 1425 (1965).
54) Bratoz, S., Durand, P.: J. Chem. Phys. *43*, 2670 (1965). — Bessis, G., Espagnet, P., Bratoz, S.: Intern. J. Quantum Chem. *3*, 205 (1969).
55) Sinanoglu, O.: Proc. Roy. Soc. (London) *A 260*, 379 (1961); Phys. Rev. *122*, 493 (1961). — McWeeny, R., Steiner, E.: Advan. Quantum. Chem. *2*, 93 (1965).
56) Diner, S., Malrieu, J. P., Clavery, P.: Theoret. Chim. Acta *13*, 118 (1969). — Diner, S., Malrieu, J. P., Jordan, M., Gilbert, M.: Theoret. Chim. Acta *15*, 100 (1969). — Jordan, F., Gilbert, M., Malrieu, J. P., Pincelli, U.: Theoret. Chim. Acta *15*, 211 (1969).
57) Kelly, H. P., Sessler, A. M.: Phys. Rev. *132*, 2091 (1963). — Kelly, H. P.: Phys. Rev. *134*, A 1450 (1964). — Kelly, H. P.: Adv. Chem. Phys. *14*, 129 (1969).
58) Freed, K. F.: Phys. Rev. *173*, 1, 24 (1968).
59) Bender, C. F., Davidson, E. R.: Phys. Rev. *183*, 23 (1969).
60) Wahl, A. C.: Advan. Quantum Chem. *4*, 261 (1971) and references given therein.
61) Meyer, W.: Intern. J. Quantum Chem. *1972*, 59.
62) This has been done to a large extent in a recent review paper. Kutzelnigg, W.: Molecular calculations including electron correlation. In: E. Clementi, ed., Selected topics in molecular physics", proceedings of a symposium held in Ludwigsburg, Germany, October 1970. Weinheim: Verlag Chemie 1972.
63) See e.g. Gelus, M., Ahlrichs, R., Staemmler, V., Kutzelnigg, W.: Theoret. Chim. Acta *21*, 63 (1971).

[64] Boys, S. F.: Rev. Mod. Phys. *32*, 300 (1960). — Edmiston, C., Ruedenberg, K.: Rev. Mod. Phys. *34*, 457 (1963). — England, W., Salmon, L. S., Ruedenberg, K.: Topics Curr. Chem. (Fortschr. Chem. Forsch.) *23*, 31 (1971).

[65] Rothenberg, S.: J. Am. Chem. Soc. *93*, 68 (1971).

[66] Kutzelnigg, W., Gelus, M.: Chem. Phys. Letters *7*, 296 (1970). — Kutzelnigg, W., Staemmler, V., Gelus, M.: Chem. Phys. Letters *13*, 496 (1972).

[67] Čížek, J.: J. Chem. Phys. *45*, 4256 (1969). — Čížek, J.: Adv. Chem. Phys. *14*, 35 (1969). — Paldus, J., Čížek, J., Shavitt, I.: Phys. Rev. *A 5*, 50 (1972).

[68] Ahlrichs, R., private communication.

[69] Hall, J. H., Marynick, D. S., Lipscomb, W. N.: Inorg. Chem. *11*, 3126 (1972).

[70] Žurawski, B., Ahlrichs, R., Kutzelnigg, W.: Chem. Phys. Letters, in press.

Received June 28, 1972

Orbital Symmetry Rules for Inorganic Reactions from Perturbation Theory

Prof. Ralph G. Pearson

Department of Chemistry, Northwestern University, Evanston, Illinois, USA

Contents

I. Introduction

The word symmetry comes from the Greek *syn* — together — and *metron* — measure — which tells us that we compare the relation of two or more things to observe symmetry. For example, we associate the word with beauty of the human form by comparing the two parts of the face and body with respect to a mirror plane bisecting our body. Symmetry is also related to the proportions of the various parts of our body with respect to each other.

Symmetry is related to order, pattern and regularity. This means that physical scientists have a large stake in symmetry because without order, pattern and regularity it would be impossible for any understanding of the physical universe to be reached. Certain areas of science use symmetry in a very detailed manner, for example, crystallography and spectroscopy. Without the use of symmetry, and the use of group theory as the mathematical tool to exploit it, it is clear that the various branches of spectroscopy would be at a very primitive level. It would not be possible, for example, to deduce the structure of molecules from their vibrational spectra.

Various theories of chemical bonding such as Hückel aromatic theory in organic chemistry and crystal field theory in inorganic chemistry are successful primarily because of the full use that they make of symmetry properties of molecules and complex ions. Arguments based on symmetry are very powerful since they usually supply answers of a yes or no variety, compared to the maybe answers of most methods used in discussing chemical bonds.

Only recently have symmetry rules been devised for predicting the course, or mechanism, of chemical reactions. The greatest credit must be given to the work of Woodward and Hoffmann [1], which stimulated the efforts of others. Of course, it is always possible to find important earlier contributions. Among these should be mentioned those of Mulliken [2], Shuler [3], and Griffing [4], all of whom used the correlation methods perfected by Woodward and Hoffmann.

In this work I wish to discuss rules obtained by a different method, using perturbation theory. Early workers using this approach were Coulson and Longuet-Higgins [5], Fukui [6], and Dewar [7]. An important step was made by Bader [8] who first brought in symmetry restrictions in an explicit way. In earlier work, the role of symmetry was implicit, but not emphasized.

Bader's paper dealt primarily with the modes of decomposition of activated complexes, a special kind of unimolecular reaction. Several early papers by Salem [9] and Pearson [10] also dealt with unimolecular reactions. Recently several papers have discussed bimolecular and termolecular reactions (Pearson [11]). Before presenting the theory in a formal sense, it may be useful to describe in words what perturbation theory eventually says about the nature of a chemical reaction, and how symmetry enters in. We start with the reactant molecules, one, two or more in number. They have a number of occupied (bonding and non-bonding), molecular orbitals, and a number of empty (non-bonding or anti-bonding) orbitals.

As the reaction progresses, electrons start to flow from some of the initially occupied molecular orbitals into some of the initially empty molecular orbitals. Actually transfer is not complete. What happens is that a whole new set of molecular orbitals are formed by mixing together of the original set. Again some will be occupied and some will be empty. Clearly the new molecular orbitals must correspond to the new bonding situation that exists for the products of the reaction.

The symmetry rule for this mixing of molecular orbitals is very simple. It can be stated as

$$\varphi_i \times \varphi_f = Q \tag{1}$$

This is read as "the direct product of the symmetry of an initially occupied MO and the symmetry of an initially empty MO which is mixed with it, must be the same as the symmetry of the reaction coordinate." Except at maximum and minimum points of the potential energy surface, the reaction coordinate is symmetric to all the symmetry operations of the reacting system. Therefore mixing can only occur between molecular orbitals of the same symmetry.

This is equivalent to saying that electrons can only flow from occupied orbitals into empty orbitals with which they have a net positive overlap. This corresponds to a very simple and graphic picture of how a chemical reaction occurs. [12] It can be applied to all chemical changes, providing we have some concept of the molecular orbitals of the system.

Obviously mixing occurs most readily between occupied orbitals of high energy and empty orbitals of low energy. Also the orbitals involved must correspond to the bonds that are made and broken in the chemical reaction. These are additional requirements which are important because they usually allow us to focus our attention on a very few of the many MO's of a complex system.

II. Theory

Consider an elementary process (concerted reaction) of any molecularity. The question is, how does symmetry enter into the variation of potential energy with changing nuclear coordinates? Group theory will first be used to obtain an exact answer to this question. Fig. 1 shows the usual adiabatic plot of potential energy *vs.* reaction coordinate. The points marked A, B, and C will be used to derive the symmetry rules since they represent characteristic features of such a plot. Any point on the diagram corresponds to some arrangement of the nuclei of the reactants. This arrangement will automatically generate a certain point group (T_d, C_{3v}, C_3, etc.). All of the symmetry properties are now contained in the irreducible representations or symmetry species of that point group.

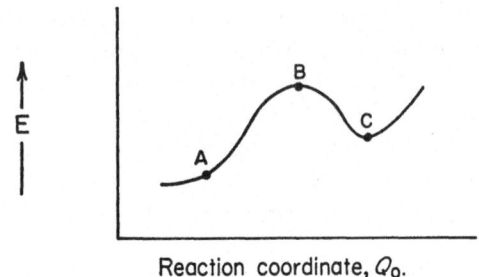

Reaction coordinate, Q_0.

Fig. 1. Plot of potential energy *vs.* reaction coordinate. Points A, B and C are referred to in text

The wave equation for the system is now assumed to be solved exactly. This gives rise to a number of eigenstates ψ_0, ψ_1, ... ψ_k, and corresponding eigenvalues E_0, E_1, ... E_k, where ψ_0 and E_0 refer to the ground electronic state. Now all the wave functions must belong to one of the symmetry species A, B, E, etc., of the point group. Indirectly then, each energy value has a symmetry label tied to it.

Any arbitrary small motion of the nuclei away from the original configuration can be analyzed as a sum of displacements corresponding to the normal modes of the pseudomolecule representing the reactants. Each of these normal modes (of vibration) belongs to one of the symmetry species of the point group.

We now use quantum mechanics in the form of perturbation theory to relate potential energy, E, to the reaction coordinate. First the Hamiltonian is expanded in a Taylor-Maclaurin series about the point Q_0, corresponding to the original configuration with Hamiltonian $\mathscr{H}_0$ (Eq. (2)). Here

$$\mathscr{H} = \mathscr{H}_0 + \left(\frac{\delta U}{\delta Q}\right) Q + \frac{1}{2} \left(\frac{\delta^2 U}{\delta^2 Q}\right) Q^2 \dots \tag{2}$$

Q represents the reaction coordinate and also the magnitude of the small displacement from Q_0. For convenience, we consider only one normal mode at a time.

Since the Hamiltonian must be invariant to all the symmetry operations of the pseudomolecule, it follows that Q and $(\partial U/\partial Q)$ have the same symmetry. Their direct product is totally symmetric. Since Q^2 is symmetric, it follows next that $(\partial^2 U/\partial^2 Q)$ is also symmetric. U is the nuclear-electronic and nuclear-nuclear potential energy. The kinetic energy of the electrons and electron-electron potential energy are not functions of the nuclear coordinates, to the first order.

The last two terms in Eq. (2) represent the perturbation. Using standard second-order perturbation theory, we now solve for the new wave functions and energies. For the ground electronic state, the energy becomes

$$E = E_0 + Q \left\langle \psi_0 \left| \frac{\delta U}{\delta Q} \right| \psi_0 \right\rangle + \frac{Q^2}{2} \left\langle \psi_0 \left| \frac{\delta^2 U}{\delta^2 Q} \right| \psi_0 \right\rangle +$$

$$Q^2 \sum_k \frac{\left[\left\langle \psi_0 \left| \frac{\delta U}{\delta Q} \right| \psi_k \right\rangle \right]^2}{E_0 - E_k} \tag{3}$$

E_0 is the energy at Q_0, the next two terms are the first-order perturbation energy, and the last term is the second-order perturbation energy. While Eq. (3) is valid only for Q very small, we can select Q_0 anywhere on Fig. 1. Hence Eq. (3) is general for the purpose of displaying symmetry properties.

The symbol $< \ldots >$ represents integration over the electron coordinates, covering all space. We can now use a group theory rule to decide whether the integrals in Eq. (3) are exactly zero or not. The rule is that the direct product of three functions must contain the totally symmetric species, or the integral over all space is zero.

Let us consider the term in (3) which is linear in Q. At any maximum or minimum in the potential energy curve, $\partial E/\partial Q = 0$ and therefore the integral must be identically zero, independent of symmetry. At all other points this term must be the dominant one, since Q is small. If ψ_0 belongs to a degenerate symmetry species (E or T), the term usually leads to the first-order Jahn-Teller effect, which removes the degeneracy. Since this is not important in the present context, we will assume that ψ_0 is nondegenerate.

Since the direct product of a nondegenerate species with itself is always totally symmetric, we derive our first symmetry rule: except at a maximum or minimum in potential energy, *all reaction coordinates belong to the totally symmetric representation*. That is, $(\partial U/\partial Q)$, and also Q, must be totally symmetric, otherwise its product with $\psi_0{}^2$ will not be symmetric and the integral will be zero. However, it *must* be nonzero for all of the rising and falling parts of Fig. 1. This means that once a reaction embarks on a particular reaction path it must stay within the same point group until it reaches an energy maximum or minimum. A totally symmetric set of nuclear motions can change bond angles and distances, but it cannot change the point group. This restriction on the point group is not as absolute as it sounds since an energy maximum may also be encountered in a normal mode orthogonal to the reaction coordinate. This then allows a nonsymmetric nuclear motion to change the point group.

We now consider point A on Fig. 1. The integral $<\psi_0|\partial U/\partial Q|\psi_0>$ has a positive value since the reaction has a positive activation energy. Instead of trying to evaluate the integral we accept that its value is the slope of Fig. 1 at the point A. The terms in Q^2 in Eq. (3) now become important. Their sum determines the curvature of the potential energy plot. For a reaction with a small activation energy, the curvature should be as small as possible (or negative).

The integral $<\psi_0|\partial^2 U/\partial^2 Q|\psi_0>$ has a nonzero value by symmetry since $(\partial^2 U/\partial^2 Q)$ is totally symmetric. Furthermore, it will be positive for all molecules. It represents the force constant which resists the movement of any set of nuclei away from an original configuration, for which ψ_0^2 is the electron density distribution. The last term in Eq. (3) represents the change in energy that results from changing the electron distribution to one more suited to the new nuclear positions determined by Q. Its value is always negative since $E_0 - E_k$ is a negative number.

This can be seen more easily if the equation for the wave function is written down from perturbation theory

$$\psi = \psi_0 + \sum_k \frac{\left\langle \psi_0 \left| \frac{\delta U}{\delta Q} \right| \psi_k \right\rangle}{E_0 - E_k} \Psi_k \tag{4}$$

The summations in (3) and (4) are over all excited states. Each excited-state wave function is mixed into the ground-state wave function by an amount shown in Eq. (4). The wave function is changed *only* because the resulting electron distribution, ψ^2, is better suited to the new nuclear positions. Salem calls the resulting decrease in energy the *relaxability* of the system along the coordinate Q. [13]

Now we can use group theory to show that only excited-state wave functions, ψ_k, which have the same symmetry as ψ_0 can mix in and lower the potential energy barrier. This follows because we have already shown that $(\partial U/\partial Q)$ must be totally symmetric. Hence the direct product of ψ_0 and ψ_k must be totally symmetric, but this requires that they have the same symmetry. We can conclude that, for a chemical reaction to occur with a reasonable activation energy, there must be low-lying excited states for the reacting system of the same symmetry as the ground state. The symmetry is related only to those symmetry elements that are conserved during the course of the reaction. Furthermore the excited states must produce changes in electron density consistent with the nuclear motions corresponding to the reaction.

Such a reaction is said to be symmetry allowed. A symmetry-forbidden reaction is simply one which has a very high activation energy because of the absence of suitable excited states for the selected reaction coordinate Q_0.

Eqs. (3) and (4) are exact, as are the symmetry rules derived from them. For practical applications, some rather drastic assumptions must now be made. One is that LCAO-MO theory will be used in place of the exact wave functions, ψ_0 and ψ_k. Since we are interested only in the symmetry properties, this creates no serious error, since MO theory has the great virtue of accurately showing the symmetries of the various electronic states.

The second assumption is more serious, since we will replace the infinite sum of excited states in (2) and (3) by only a few lowest lying states. This procedure will work because we are not trying to evaluate the sum but only to decide if it has a substantial value. It can be shown [8] that the various states contributing to (2) and (3) fall off very rapidly as the difference $|E_0 - E_k|$, becomes large. This is because the integral $<\psi_0|\partial U/\partial Q|\psi_k>$ decreases very rapidly for two wave functions of quite different energy.

Accordingly we use MO theory to represent the ground and excited states that are needed. The symmetry of $\psi_0\psi_k$ is replaced by $\varphi_i\varphi_f$, where φ_i is the occupied MO in the ground state and φ_f is the MO occupied in its place in the excited state. Positions of special importance are occupied by the highest occupied and lowest unoccupied molecular orbitals, since excitation of an electron from HOMO to LUMO defines the lowest excited state.

III. The Bond Symmetry Rule

We now go to a consideration of points B and C in Fig. 1. B refers to an activated complex and C to a single molecular species, which is unstable with respect to isomerization, or breakdown to other products. In either case the theory is changed somewhat from that of the previous section.

The term linear in Q in Eq. (2) now vanishes, since we are at an extremum in the potential energy plot. As before, the first quadratic term is positive, and the second one is negative. Clearly at a maximum, point B, the second term is larger than the first. At a minimum, point C, the first term dominates, but the magnitude of the second term determines whether we lie in a deep potential well or a shallow one.

Again the existence of low-lying states, ψ_k, of the correct symmetry to match with ψ_0 is critical. Now there is no restriction on the reaction coordinate which forces it to be totally symmetric. However, ψ_0, $(\partial U/\partial Q)$, and ψ_k *are still bound by the symmetry requirement that their direct product must contain the totally symmetric representation.*

If we consider rather symmetrical molecules to begin with, it will usually be found that the reaction coordinate, and $(\partial U/\partial Q)$, are nonsymmetric. The reason for this is that maximum and minimum potential energies are usually found for nuclear arrangements with a high degree of symmetry. Any disturbance of the nuclear positions will now reduce the symmetry. However, this corresponds to a change in the point group, which can only come about by a nonsymmetric vibrational mode.

Conversely, it may be pointed out that a number of point groups depend upon a unique value of Q_0 in Fig. 1. For example, a tetrahedral molecule has uniquely determined bond angles. All such cases must correspond to either maxima or minima in Fig. 1 if the reaction coordinate is taken as one which destroys any element of symmetry.

In molecular orbital theory the product $\psi_0\psi_k$ is again replaced by $\varphi_i\varphi_f$, where both the occupied and empty MO's must be in the same molecule. Electron transfer from φ_i to φ_f results in a shift in charge density in the molecule. Electron density increases in the regions where φ_i and φ_f have the same sign (positive overlap) and decreases where they have opposite signs (negative overlap). The positively charged nuclei then move in the direction of increased electron density. The motion of the nuclei defines a reaction coordinate. The symmetry of Q is the same as that of the product $\varphi_i \times \varphi_f$.

The size of the energy gap between φ_i and φ_f is critical. A small gap means an unstable structure, unless no vibrational mode of the right symmetry exists for the molecule capable of changing its structure. A large energy gap between the HOMO and the LUMO means a stable molecular structure. Reactions can occur but only with an activation energy.

For an activated complex (point B) there must *necessarily* be at least one excited state of low energy. The symmetry of this state and the ground state then determines the mode of decomposition of the activated complex. This was the subject of the first application of Eq. (3) to chemical reactions by Bader[8].

When a molecule lies in a shallow potential well (point C), the activation energy for unimolecular change is small. In this case we can again expect a low-lying excited state. The symmetry of this state and the ground state will determine the preferred reaction of the unstable molecule. For a series of similar molecules, we expect a correlation between the position of the absorption bands in the visible-uv spectrum and the stability.

For molecules which lie in deep potential wells, it may not be the LUMO which is important. The reason is that, since a high activation energy is required, higher lying states may be utilized. It is also difficult to place the higher excited states of a molecule in correct order. Nevertheless the symmetry rules may still be of great help in selecting the reaction path.

Suppose we know that a certain unimolecular reaction occurs, but do not know the detailed mechanism. Certain bonds must be made and broken during the reactions. The bonds can then select φ_i and φ_f. These MO's in turn will fix the symmetry of the reaction coordinate, Q. The only requirement is a knowledge of the symmetries of the MO's which relate to the bonds that are affected.

Clearly chemists are much more familiar with chemical bonds and lone pairs of electrons than they are with molecular orbitals. For this reason it is convenient to transform the usual (canonical) MO's of reactants and products into MO's which correspond more closely to chemical bonds. The canonical MO's are the solutions of the secular determinant which follows from use of the variational method of quantum mechanics. In the LCAO method, these MO's are made up chiefly of the valence shell AO's of the atoms involved.

As reaction occurs to form products, the same valence shell AO's are used to form a new set of MO's, exactly the same number as before, but differing in composition and in bonding characteristics. By perturbation theory the new MO's are formed from the old by a mixing of the originally empty with the originally filled ones. Perturbation theory can be applied at each point along the reaction coordinate to make this process continuous. Only orbitals of the same symmetry can mix continuously.

Some of the new MO's will differ but little from the original MO's because they correspond to similar bonding situations. The greatest changes will occur in the orbitals that correspond most closely to the changes in bonding. While often these orbitals can be identified, we wish to convert the canonical MO's of reactant and products into more localized MO's corresponding to definite bonds between a limited number of atoms or to lone pairs isolated on a single atom. Such bonding orbitals are formed by linear combinations of the canonical orbitals of the same symmetry. Only the occupied MO's of each molecule can be combined to give the occupied bonding orbitals. The empty MO's combine to give the anti-bonding partners.

These bonding orbitals are not the same as the localized orbitals usually considered as equivalent to chemical bonds [15] since the latter are mixtures of MO's of different symmetries. However, they can be directly related to the usual chemical bonds of the molecule. Thompson [15] has shown how the bonds and lone pairs of a molecule can be used as a basis set for symmetry classification. Thus each unique bond in a molecule has a definite symmetry label. In the case of two or more identical bonds, symmetry adapted linear combinations of the bonds must be formed. Each of the symmetry adapted bonds will correspond to one of the bonding orbitals described above.

To see how this useful procedure works, let us consider the bonds and the molecular orbitals of the water molecule. Fig. 2 shows a Lewis diagram of H_2O, with two lines for OH bonds and two pairs of dots for lone pairs of electrons. The point group is C_{2v} and the lone pairs are placed above and below the plane of the molecule. For a discussion of the stereochemistry of lone pairs, the reader should consult the papers by Gillespie [16].

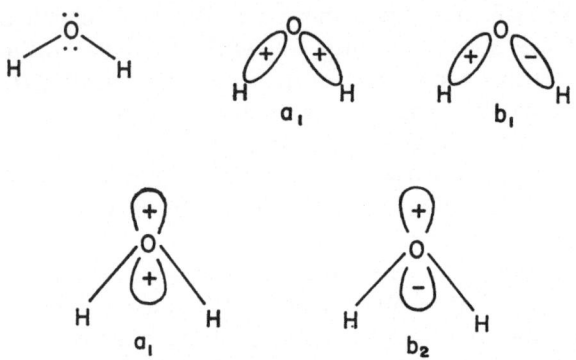

Fig. 2. Lewis diagram for H_2O, and the symmetry adapted bond orbitals that can be generated from it

The two bonds must be considered together. Their sum and difference are symmetry adapted linear combinations. These are now shown in Fig. 2 as localized MO's corresponding to bonds between O and H. In the C_{2v} point group, the symmetries are clearly a_1 and b_1. The sum and difference of lone pair orbitals concentrated on O would correspond to a_1 and b_2 symmetries. For convenience Fig. 3 shows the symmetry species of the C_{2v} point group. Only the behavior with respect to two mirror planes is necessary to classify all four species. There is also a two-fold axis at the intersection of the two planes.

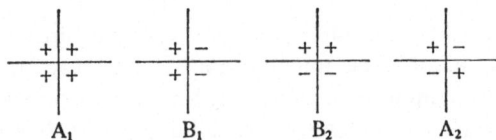

Fig. 3. Symmetry species for the C_{2v} point group. Behavior with respect to two mirror planes is shown. A two-fold axis is perpendicular at the intersection of the two planes

The canonical MO's of water in order of increasing energy give the configuration

$$(1a_1)^2(2a_1)^2(1b_1)^2(3a_1)^2(1b_2)^2.$$

The $1a_1$ orbital is the $1s$ orbital of oxygen and is not part of the valence shell. The composition of the other orbitals is shown in Table 1. The b_1 orbital corresponds to the difference of the two OH bond orbitals. The b_2 orbital is the antisymmetric lone pair orbital.

Table 1. Coefficients of atomic orbitals in molecular orbitals of H_2O[1]

$2a_1$	$0.845\ (O_s)$	$+$	$0.133\ (O_z)$	$+$	$0.178\ (H_1s + H_1s)$
$1b_1$	$0.543\ (O_y)$	$+$	$0.776\ (H_1s - H_1s)$		
$3a_1$	$-0.460\ (O_s)$	$+$	$0.828\ (O_z)$	$+$	$0.334\ (H_1s + H_1s)$
$1b_2$	$1.000\ (O_x)$				

[1] From Ellison, F. O., Shull, H.: J. Chem. Phys. *23*, 2348 (1955).

We now take linear combinations of the $2a_1$ and $3a_1$ orbitals, which are the only ones of the same symmetry. The intent would be to get one orbital concentrated on oxygen, representing the second lone pair, and devoid of bonding characteristics. The second orbital would then be a strongly bonding orbital for both O—H bonds. The detailed composition of these orbitals does not concern us. The important conclusion is that we can assign definite symmetries to the bonds and lone pairs of the water molecule.

The same thing can be done for all other molecules for which simple Lewis diagrams can be drawn. For molecules such as benzene or sulfur dioxide, where this is not possible, we must treat certain bonds as delocalized. In this case we can still get the symmetries by a simple Hückel type of calculation. [15]

We now return to considering motion along the reaction coordinate which mixes the occupied and empty MO's of the same symmetry. We focus our attention on the bonds to be broken which clearly select certain φ_i. The bonds to be made select the orbitals φ_f, since the only way we can create *new* bonds is by mixing φ_i and φ_f. With respect to the elements of symmetry that are conserved, φ_i and φ_f must be of the same symmetry. More important, the new bonding orbital that is formed *must* be of the same symmetry as the bonding orbital that was destroyed.

We have accordingly derived a simple and generally applicable symmetry rule: *a reaction is allowed if the symmetry of the bonds that are made is the same as the symmetry of the bonds that are broken.*[17] The symmetry is related only to those symmetry elements that are conserved in going from reactants to products. The rule is applicable to reactions of any molecularity.

We can reach the same conclusions by using orbital correlation arguments. If the symmetries of the bonds that are broken and the symmetries of the bonds that are made always match up in pairs, the non-crossing rule will then guarantee that none of the corresponding pairs of orbitals will cross. If they do not cross, leading to a hypothetical excited state product[a], the reaction is allowed by the orbital correlation procedure.

Fig. 4 shows schematically how canonical MO's are transferred to bond orbitals, and how the bond orbitals of the reactants transform to those of the products. A bonding orbital is shown interacting with an anti-bonding orbital of the same symmetry. The result is the new bond orbital characteristic of the products.

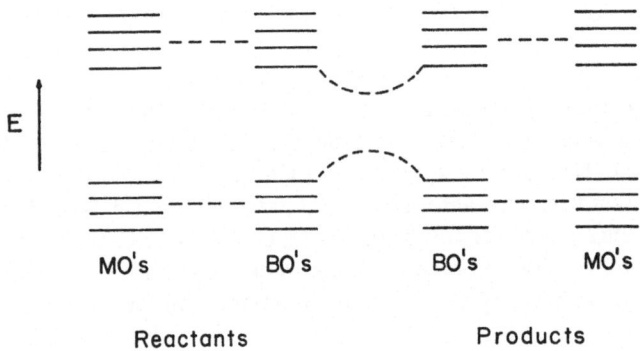

Fig. 4. Transformation of canonical MO's into bond orbitals (BO's) for reactants and products. Lower orbitals are occupied and higher ones are empty in each case. The central dashed lines show interaction of a bonding orbital with an anti-bonding orbital of the same symmetry

The above bond symmetry rule suggests immediately that certain reactions are free from orbital symmetry restrictions. These are reactions in which bonds are only broken, but none made, or in which bonds are only made, but none broken. Typical examples would be dissociation of a molecule into atoms or radicals, and the reverse process of radical recombination.

---

a) It should perhaps be emphasized again that the crossing implied is only an intended crossing. Configuration interaction will usually prevent an actual crossing.

Such reactions are indeed relatively free of orbital symmetry restrictions, but not entirely so. The electrons of a bond that is broken must still end up somewhere. The above rule says that they must end up in orbitals of the same symmetry as the bond that they originated from. For example, as HCl dissociaties into atoms, the electrons of the σ bond must remain in atomic orbitals of σ symmetry. Essentially this restriction, plus the spin conservation rule, is the basis of the Wigner-Witmer rules.[18] In the same way, a bond that is formed must conserve the symmetry of the original orbitals of the constituent electrons. Additionally, bonding can only occur by the combination of atomic orbitals of the same symmetry (net positive overlap).

IV. Examples of Bond Symmetry Rule

We start by considering possible reactions of diatomic molecules with each other, *e.g.*,

$$H_2 + I_2 = 2\,HI \tag{5}$$

$$I_2 + Cl_2 = 2\,ICl \tag{6}$$

$$N_2 + O_2 = 2\,NO \tag{7}$$

The simplest mechanism that we can visualize for such reactions would involve a four-center transition state formed by a broadside, bimolecular collision.

$$
\begin{array}{ccc}
\begin{matrix} I & & Cl \\ | & + & | \\ I & & Cl \end{matrix}
& \longrightarrow &
\begin{matrix} I\text{-}\text{-}Cl \\ | \quad | \\ I\text{-}\text{-}Cl \end{matrix}
\quad \longrightarrow \quad
\begin{matrix} I\text{--}Cl \\ + \\ I\text{--}Cl \end{matrix}
\end{array}
\tag{8}
$$

$$\qquad 2\,a_1 \qquad\qquad\qquad a_1 + b_1$$

In the C_{2v} point group, the symmetry elements of which are conserved during the reaction, we start with two unique bonds of a_1 symmetry. Two identical bonds are formed which must be considered together. Their sum and difference are of a_1 and b_1 symmetry, respectively. Hence the reaction is forbidden by orbital symmetry. The same conclusion would be drawn for all other reactions of homonuclear diatomic molecules with each other.

Indeed we now know that all such reactions which have been carefully studied do *not* go by the simple mechanism of Eq. (8).[10b,10c] The reactions of the halogens with each other occur by free atom chain reactions in the gas phase [19], or by slow and mysterious mechanisms in solution.[20]

The same conclusion of forbiddenness results if we consider addition reactions of diatomic molecules, such as,

$$
\begin{array}{ccc}
\mathrm{N}\!\!\equiv\!\!\mathrm{N} & & \mathrm{N}\!\!=\!\!\mathrm{N} \\
+ & \longrightarrow & |\quad| \\
\mathrm{H}\!\!-\!\!\mathrm{H} & & \mathrm{H}\quad\mathrm{H}
\end{array}
\tag{9}
$$

$$2a_1 \qquad\qquad a_1 + b_1$$

We need not restrict ourselves to diatomic molecules since in symmetry terms, reaction (10) is the same as (9).

$$
\tag{10}
$$

$$2a_1 \qquad\qquad a_1 + b_1$$

In general we can conclude that all reactions are forbidden in which we take bonds that are originally written as up and down and convert them to bonds that are written as left to right.

We abandon the broadside collision with its four-center transition state and consider other possible orientations, such as end-on.

$$
\mathrm{I-H+H-I} \longrightarrow \mathrm{I\text{-}\text{-}H\text{-}\text{-}H\text{-}\text{-}I} \longrightarrow \mathrm{I+H_2+I} \tag{11}
$$

The point group is now $D_{\infty h}$. The bonds that are broken are σ_g and σ_u for the sum and difference of the two H—I bonds. The new bond formed is of σ_g symmetry, and two electrons are left in the p_σ orbitals of the two iodine atoms. These can be added to give σ_u symmetry. The reaction is allowed.

Reaction (11) illustrates what generally happens if we abandon the highly symmetric broadside collision mechanism. A less symmetric collision will circumvent the orbital symmetry restriction, but it will necessitate the formation of high energy intermediates. In this case an energetic price must be paid since two free iodine atoms are formed. Even so it appears that reaction (11), or some less symmetrical version of it, is what actually occurs. Semi-empirical calculations, while not completely to be trusted, indicate that linear IH_2I is more stable than trapezoidal H_2I_2 (Minn and Hanratty [21]).

It is important to realize that this calculated result is what must necessarily be true if the orbital symmetry arguments are valid. That is, the activated complex for the allowed path must lie at a lower energy than that of the forbidden path. Symmetry barriers are not mysterious obstructions put in the way of otherwise energetically favorable transition states. They instead directly affect the transition state energy.

Eq. (10) shows that concerted *cis* addition of hydrogen to ethylene is forbidden. However, concerted *trans* addition is allowed.

$$\overset{H}{\underset{H}{}}C \!\!=\!\! C \quad \longrightarrow \quad \overset{H}{\underset{H}{C - C}} \tag{12}$$

This kind of addition is called *antarafacial* by Woodward and Hoffmann[1]. The *cis* addition is called *suprafacial*.

Fig. 5 shows why the former process is allowed, using the bond symmetry criterion. The point group becomes C_2, with a single two-fold axis passing through the center of both molecules. The H—H bond is of a symmetry, and the C—C π bond is of b symmetry. The sum and difference of the two C—H bonds formed are a and b respectively. The ethane that is formed by *antarafacial* addition is quite strained, with abnormal bond angles. This means an energy price must again be paid to make a forbidden reaction allowed.

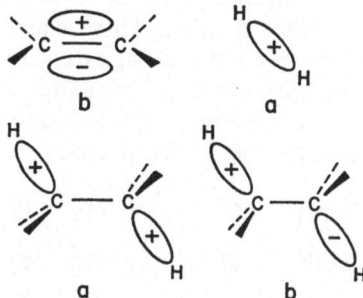

Fig. 5. The symmetries of the C—C π bond and the H—H σ bond in the C_2 point group. The symmetries of two equivalent C—H σ bonds in C_2

Notice that the skew approach of two diatomic molecules will not always make the reaction allowed.

$$\overset{H}{\underset{H}{I - I}} \quad \longrightarrow \quad \underset{\underset{H}{\overset{+}{}}{I}}{I}\overset{H}{} \tag{13}$$

In the C_2 point group, the σ bonds of both I_2 and H_2 are of a type. The two H—I bonds are of $a+b$ symmetry. The reaction is still forbidden.

The reactions of chlorine gas with either dry sulfur dioxide, or with carbon monoxide are remarkably slow. In both cases reaction only occurs at high temperature, and by a complex process involving free chlorine atoms (Bamford and Tipper [22]). Simple concerted mechanisms can be imagined in both cases.

$$\text{(14)}$$

$$a_1 \qquad a_1 \qquad\qquad a+b_2$$

$$\text{(15)}$$

$$a_1 \qquad a_1 \qquad\qquad a_1+b_1$$

The point group in both cases is C_{2v}. The bonds that are made in SO_2Cl_2 are of a_1 and b_2 symmetry (the sum is a_1, the difference is b_2). The Cl—Cl bond that is broken is of a_1 symmetry. We now must find the highest energy lone pair of electrons in sulfur dioxide. These electrons are the ones that are involved when SO_2 acts as a reducing agent, as in this case. The HOMO of SO_2 in fact is of a_1 type, concentrated on sulfur as shown. There are *no* electrons in the orbital of b_2 symmetry. The reaction (14) is thus forbidden.

Compare this situation to that of the reaction

$$\text{(16)}$$

$$b_2 \qquad a_1 \qquad\qquad a_1+b_2$$

there are two more electrons in the valence shell of SCl_2 than in SO_2. These do go into the lone pair orbital of sulfur of b_2 symmetry, as shown. The reaction is allowed, and indeed occurs instantaneously, even at 0 °C.

Similarly in carbon monoxide, the most loosely held electrons are in an orbital of a_1 type, concentrated on carbon as shown in reaction (15). Again the mechanism is a forbidden one, since one bond of different symmetry, b_1, must be formed.

There may be some uneasiness in the reader's mind at this point because the selection of the HOMO seems rather arbitrary. However a rigorous check is possible. We examine *all* the valence shell MO's of CO and Cl_2

and compare them to those fo $COCl_2$. They will not correspond (Mahan [23]). This is most easily done by drawing a Lewis diagram of each molecule.

$$
\begin{array}{ccc}
Cl & & Cl \\
| + :C\!=\!\ddot{O} & \longrightarrow & \rangle C\!=\!\ddot{O} \\
Cl & & Cl
\end{array} \tag{17}
$$

$$a_1 \quad 3a_1 + b_1 + b_2 \quad 3a_1 + 2b_1 + b_2$$

Note that the π bond of $C=O$ is, and remains, of b_2 symmetry. The lone pairs on oxygen are, and remain of $a_1 + b_1$ symmetry.

Concerted *cis* addition of Cl_2 to an olefin is a forbidden process.

$$
\begin{array}{ccc}
\rangle C\!=\!C\langle & & \rangle \dot{C}\!-\!\dot{C}\langle \\
+ & \longrightarrow & Cl \quad Cl \\
Cl\!-\!Cl & &
\end{array} \tag{18}
$$

$$2a_1 \qquad\qquad a_1 + b_1$$

However concerted *cis* addition of two chlorine atoms from molecules such as $PbCl_4$ or $SbCl_5$ is allowed.

$$
\begin{array}{ccccc}
Cl\diagdown \quad \diagup Cl & & C & & Cl\diagdown \\
\quad Pb & + & \| & \longrightarrow & \quad Pb \oplus \\
Cl\diagup \quad \diagdown Cl & & C & & Cl\diagup
\end{array} + \quad \begin{array}{c} Cl\diagdown \; C \\ | \\ Cl\diagup C \end{array} \tag{19}
$$

$$a_1 + b_1 \qquad a_1 \qquad\qquad a_1 \qquad a_1 + b_1$$

The lone pair of electrons in $PbCl_2$ is in an orbital of a_1 symmetry, primarily an s orbital on lead. It is interesting to note that $SbCl_5$ does yield *cis*-dichloro products from olefins in a concerted process (Uemura, Sasaki and Okano [24]). Following the arguments of Hoffmann, Howell and Muetterties [25], the chlorine atoms must be removed, one from an axial and one from an equatorial position in $SbCl_5$.

It has been claimed by Dewar [26] that $C_6H_5ICl_2$ also adds chlorine *cis* to olefins in a concerted process.

$$C_6H_5ICl_2 + C_2H_4 \longrightarrow C_6H_5I + C_2H_4Cl_2 \tag{20}$$

If we examine this reaction from the viewpoint of bond symmetry, we find that it is forbidden.

$$2a_1 + b_1 + b_2 \qquad a_1 \qquad\qquad a_1 + b_1 + b_2 \qquad a_1 + b_1 \tag{21}$$

The two lone pairs of electrons in $ArICl_2$ lie in the same plane, perpendicular to the plane of the page. Their sum is of a_1 symmetry and their difference is b_1. The two I—Cl bonds give $a_1 + b_2$ species. In ArI, the three lone pairs on iodine must have the symmetries of the three p orbitals on iodine. These are a_1, b_1 and b_2. Indeed on examining the geometry of $C_6H_5ICl_2$ (Archer and van Schalkwyk [27]) it is found that the two chlorine atoms are 4.90 A apart, which makes a concerted addition to an olefin bond quite impossible in any case.

The concerted addition of MnO_4^- or OsO_4 to the olefinic double bond is usually considered to be the first step in the hydroxylation reaction. A number of such reactions have been discussed by Littler. [28]

$$a_1 \qquad\qquad a_1 + b_2 \qquad\qquad 2a_1 + b_2 \tag{22}$$

From the viewpoint of bonds and lone pairs, the most important change is that Os(VIII) is reduced to Os(VI). Two electrons are transferred from the olefinic double bond (of a_1 symmetry) to the metal ion. Now the tetra-coordinated Os(VI) product, while formally of C_{2v} symmetry, is essentially a tetrahedral complex. In that case the $d_{x^2-y^2}$ and d_{z^2} orbitals are lower in energy than d_{xy}, d_{xz} and d_{yz}. Hence the electrons will appear in one or the other of the former orbitals, both of which are of a_1 symmetry in C_{2v}. The symmetries match and the reaction is allowed.

We now consider if OsO_4 or MnO_4^- can attack a saturated hydrocarbon, causing oxidation to an olefin by removing two hydrogen atoms in a concerted process.

$$a_1 + b_2 \qquad\qquad a_1 + b_2 \qquad\qquad a_1 \qquad\qquad 2a_1 + b_2 \tag{23}$$

In this case the symmetries of the bonds and lone pairs involved do not match. The manganese(VII) is reduced to manganese(V) and a pair of electrons must appear again in a d orbital of a_1 symmetry. The process is forbidden.

As a final example we consider a reaction in which highly delocalized bonds are involved. In this case we must use the molecular orbitals corresponding to the delocalized bonds. A case in point would be the conversion of Dewar benzene to benzene, a strongly forbidden process (Woodward and Hoffmann [1]).

Fig. 6 presents the necessary analysis. The elements of the C_{2v} point group are conserved in a concerted ring opening reaction. The sum and difference of the two π bonds of Dewar benzene are of $a_1 + b_1$ symmetry. The long carbon-carbon σ bond is a_1. The three occupied π orbitals of benzene are of a_{2u} and e_{1g} species in the D_{6h} point group of the molecule. However we are interested only in their symmetry in C_{2v}. As shown in Fig. 6 the three occupied MO's of π type are a_1, b_2 and b_1. Correlation with the bonds of Dewar benzene is not possible.

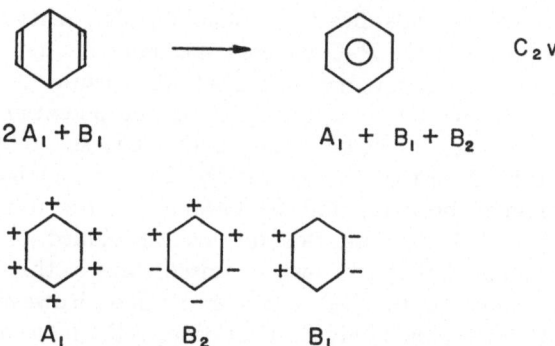

Fig. 6. The symmetries of the bond orbitals of Dewar benzene and benzene in the C_{2v} point group

V. Predicting the Reaction Coordinate

In the previous examples the procedure was to select a reaction coordinate and to see if a given reaction was compatible with it. It should be possible to reverse the procedure. That is, to select a reaction by focusing on the bonds that are to broken and made, and then let these bonds determine the reaction coordinate.

This procedure is feasible only for unimolecular reactions, since only in this case do we have a definite starting configuration. Let us consider a simple unimolecular reaction, the decomposition of water.

$$H_2O \longrightarrow H + OH \tag{24}$$

This reaction proceeds, according to perturbation theory, by electrons moving from occupied bonding, or non-bonding, orbitals into empty, anti-bonding orbitals. Alternatively the anti-bonding orbitals are mixed with the bonding orbitals to form new molecular orbitals.

The anti-bonding MO's from the valence shell are of a_1 and b_1 type. Just as we can assign symmetry labels to bond orbitals, so we can have anti-bond orbitals of definite symmetry. For water we must take the sum and difference of the two anti-bonding orbitals to be consistent with C_{2v} symmetry.

The only way we can have an asymmetric dissociation of the water molecule, as in (24), is to promote electrons from an a_1 bond orbital into a b_1 anti-bond orbital, or from a b_1 bond orbital into an a_1 anti-bond orbital. In both cases the symmetry of the reaction coordinate is $A_1 \times B_1 = B_1$, according to Eq. (1). This is, of course, the asymmetric stretch of the water molecule which corresponds directly to reaction (24). After a slight distortion the point group of the H_2O molecule becomes C_s.

This reduction in symmetry is important because as earlier noted, unless we are at a maximum or minimum in the potential energy curve, the reaction coordinate must be totally symmetric and mixing can occur only between orbitals of the same symmetry. In C_s, which has only a single plane of symmetry, both A_1 and B_1 become A', the totally symmetric representation, and the necessary conditions are fulfilled.

It should be noted that the lowest excited state of the water molecule is B_2, corresponding to the $(b_2) \rightarrow (a_1^*)$ excitation. However a B_2 or A_2 motion of H_2O corresponds only to a rotation of the molecule (see Fig. 7). Also the second excited state of the water molecule is of A_1 symmetry (Herzberg [29]). But this corresponds to a reaction coordinate which is symmetric, i.e., to the reaction

$$H_2O \longrightarrow 2H + O \tag{25}$$

It is clear that for reactions which require appreciable activation energies that excited states which are quite high in energy maybe the important ones in Eq. (4). The lowest excited state cannot be used as a guide to reactivity (Salem [13], Pearson [11a]).

The previous example is a rather trivial one because of the simplicity of a triatomic molecule. More complex molecules offer a wide range of possible reaction coordinates. The ring opening reactions of organic molecules provide useful examples. As Woodward and Hoffmann [1] pointed out, such ring openings can occur in two distinct ways, called disrotatory and conrotatory. Fig. 7 shows that these are respectively of B_2 and A_2 symmetry in the C_{2v} point group.

We deduce that ring opening in cyclobutene occurs in a conrotatory fashion by considering first the symmetry of the bonds we wish to break. These are a C—C σ bond of a_1 symmetry and a C—C π bond of b_2 symmetry.

We can destroy a bond in one of two ways: by removing electrons from the bonding orbital, or by adding electrons to the corresponding anti-bonding orbital. The anti-bonding orbitals in this case are a σ^* orbital of b_1 type and a π^* orbital of a_2 type.

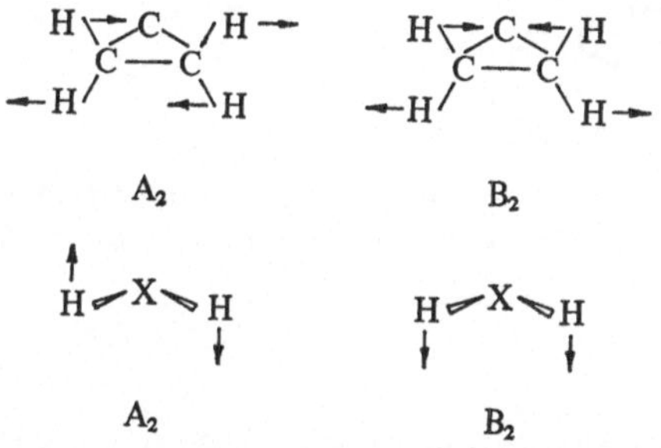

Fig. 7. Above: Symmetries of conrotatory (A_2) and disrotatory (B_2) nuclear motions in the C_{2v} point group
Below: Nuclear motions of A_2 and B_2 symmetry for atoms lying in a plane C_{2v}

Fig. 8 shows that electron transfer from σ to π^*, or from π to σ^* both result in a reaction coordinate of A_2 symmetry (using Eq. (1)). This is the conrotatory mode. The alternative choice of mixing σ with σ^* and π with π^* gives a reaction coordinate of $A_1 \times B_1 = A_2 \times B_2 = B_1$ symmetry. This motion is an asymmetric distortion in the plane of the molecule which cannot lead to symmetric ring opening.

After twisting of the methylene groups begins (preceded or accompanied by stretching of the carbon-carbon σ bond), the point group becomes C_2, only a two-fold axis being preserved. In this lower symmetry both A_1 and A_2 become A, and B_1 and B_2 both become B. This is a necessary requirement. As conrotatory twisting continues the two A orbitals combine and the two B orbitals combine to form the two new molecular orbitals which are the π orbitals of butadiene.

Notice that the alternative choice of $\sigma - \sigma^*$, or $\pi - \pi^*$, mixing is not allowed because of different symmetries, both in C_2 and in the C_s point

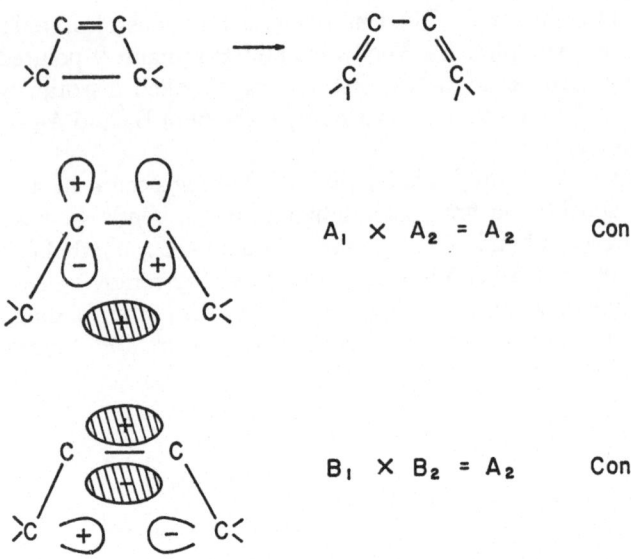

Fig. 8. Mixing of σ bonding orbital and π^* anti-bonding orbital or of π bonding orbital and σ^* anti-bonding orbital both give A_2 (conrotatory) symmetry. Filled orbitals are shaded

group generated by a disrotatory motion. Also in C_s, A_1 becomes A' and A_2 becomes A''. Thus a disrotatory mode is incompatible with breaking the selected bonds.

As an example of a reaction involving a transition metal ion, let us try to predict the behavior of a molecule which has been postulated as an intermediate in the olefin disproportionation reaction catalyzed by transition metal complexes. This reaction has been discussed by Hughes [230a)], Mango and Schachtschneider [30b)], Lewandos and Pettit [30c)].

$$\begin{array}{c} \overset{a}{}\!\!\diagdown\!\!\diagup^{c} \\ \diagup\!\!\diagdown \\ ^{b}\diagup\diagdown_{d} \end{array} \xrightarrow{\text{cat.}} \begin{array}{c} \overset{a}{}\!\!\diagdown\diagup^{c} \\ =\!\!=\!\!= \\ ^{b}\diagup\diagdown_{d} \end{array} \qquad (26)$$

There is considerable evidence that an intermediate such as $Mo(CO)_2$ (olefin)$_2$ is involved. A planar structure of C_{2v} symmetry will be assumed. However the conclusions would be the same if the structure were pseudo-tetrahedral, or even if the intermediate were $Mo(CO)_3$(olefin)$_2$ or $Mo(CO)_4$-(olefin)$_2$, provided the point group was C_{2v}.

By adding and subtracting the π orbitals of the two olefins (pointing in towards the metal atom) we generate orbitals of a_1 and b_1, species. The sum and difference of the π^* orbitals similarly gives b_2 and a_2 species.

The metal d orbitals in C_{2v} symmetry transform as $2a_1$, b_1, b_2, a_2. Which orbitals are of each symmetry depends on our choice of x, y and z axes. Olefin orbitals and metal orbitals of the same symmetry will combine to give a bonding scheme which has been discussed previously.[30] The details do not need to concern us.

The only necessary information is that both filled (bonding) and empty (anti-bonding) MO's exist of a_1, b_1, a_2 ,and b_2 symmetry. We will assume, with previous workers, that a key step is the formation of a complexed cyclobutane-like molecule, which can then decompose to the original complex or a new complex with the dismuted olefin molecules as ligands. These steps are all allowed by either orbital correlation, or perturbation theory.

Fig. 9 shows, however, an unpleasant development; the new complex must have pseudo-tetrahedral structure, as pointed out by Caldow and MacGregor [31]. While planar and tetrahedral complexes sometimes interconvert readily, the activation energy is usually greater than 10 kcal/mole, according to studies by Pignolet, Horrocks and Holm [32]. The olefin disproportionation reaction has an experimental activation energy of only 6—7 kcal.

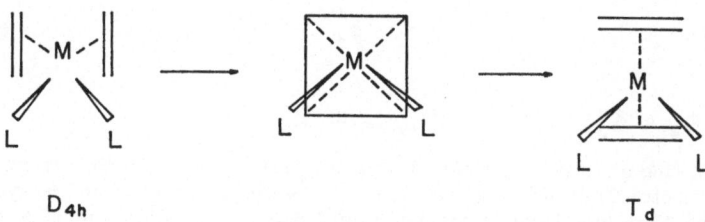

Fig. 9. Olefin disproportion with C_{2v} point group fixed leads to tetrahedral dilemma

At this point we can gain additional information from perturbation theory, not easily available from orbital correlation diagrams. If we wish to join two olefin molecules together to form cyclobutane, it is clearly necessary to transfer electron density from a filled orbital of b_1 symmetry, to an empty orbital of b_2 symmetry (see Fig. 10). This is the only way in which we can break the carbon-carbon π bonds and convert them into suitable σ bonds. Mango and Schachtschneider [30b] originally accomplished the necessary change by moving electrons from a filled b_1 orbital (concentrated on the olefins) into the empty b_1^* orbital (concentrated on the metal), and from the filled b_2 orbital (concentrated on the metal) into the empty b_2^* orbital (concentrated on the olefins).

This is perfectly permissible, but the symmetry rules then demand that the reaction coordinate be of A_1 symmetry. Thus the C_{2v} point group must be maintained throughout. This in turn means that we encounter the square planar-tetrahedral dilemma of Fig. 9.

There is another procedure possible and that is to move electrons from the filled orbital of b_1 type directly into the empty b_2^* orbital. Since $B_1 \times B_2 = A_2$, the reaction coordinate is now of A_2 type. An A_2 motion reduces the symmetry from C_{2v} to C_2. In this lower point group both b_1 and b_2^* become of b symmetry, and they can freely mix. Fig. 10 shows what happens when a b_1 orbital is mixed with a b_2 orbital. The π and π^* orbitals of the two olefin molecules are shown, and also half of two d orbitals of the metal which are of the b_1 and b_2 symmetry.

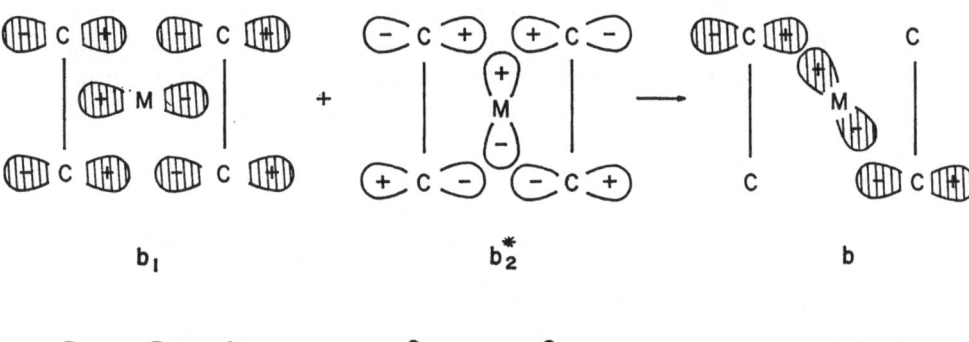

$$B_1 \times B_2 = A_2 \qquad C_{2v} \longrightarrow C_2$$

Fig. 10. Mixing of filled b_1 orbital with empty b_2^* orbital gives rise to twisting motion of A_2 symmetry. Half of two d orbitals on the central metal of b_1 and b_2 symmetry are shown. Note how bonding changes from between the carbon atoms to towards the carbon atoms

An A_2 vibrational mode corresponds to a rotation of the two parts of the molecule with respect to each other. There will be a cyclic twist of electron density in one direction in the region of the four carbon atoms of the olefin ligands, and a compensating electron movement on the metal and remaining ligands. The nuclei will then move to follow the changed electron density. The structure of the complex becomes as shown in Fig. 11. The advantage of this structure is that it can go on to form dismuted olefin without passing into the high energy tetrahedral structure. While the metal to carbon bonding must be weakened in the cyclobutane-like intermediate, there is little chance that free cyclobutane will be released (none is found). This would correspond to a one-step change in coordination number of two units, a situation not found in coordination chemistry, unless accompanied by oxidation-reduction.

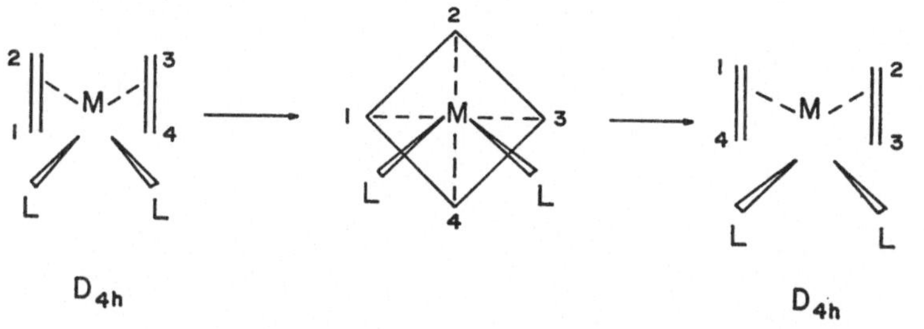

Fig. 11. Twisting of cyclobutane ring avoids the tetrahedral dilemma

VI. Reactions of No Symmetry

A. General Considerations

In the great majority of molecules, of course, no elements of symmetry exist at all. The point group is C_1. Similarly, in most reactions no useful elements of symmetry are conserved. How can one then decide which reactions are forbidden and which are allowed? Actually the situation is very analogous to that of forbidden transitions in electronic spectroscopy. Reduction in symmetry can make such transitions allowed. However there will still be an inherent forbiddenness which shown up as a low intensity of absorption.

In the same way a reduction in symmetry can make a forbidden reaction partly allowed, but still one that has a high activation energy. Table 2 shows some rate data for isomerization of substituted Dewar benzenes to the corresponding benzene derivatives. We see that a single chlorine or fluorine substituent increases the rate of isomerization markedly.

Such a substituent lowers the symmetry from C_{2v} to C_s. In this point group the original bonds are $2a'$ and a'', and the final bonds are also $2a'$ and a''. From Fig. 6 one can see that the troublesome b_2 orbital, which causes the forbiddenness in Dewar benzene, becomes a', just as the a_1 bond orbital becomes a'. The reaction is formally allowed. Adding two chlorine atoms returns the symmetry to C_{2v} again, and the rate falls off drastically. Both a symmetry factor and an electronic factor of some other kind are manifest. Nevertheless the "allowed" reactions are still remarkably slow and have very substantial activation energies. The reactions are all very exothermic to the extent of about 60 kcal. There is no obvious reason why an activation energy should exist at all, if it were not for the lingering effect of the symmetry forbiddenness of the symmetrical parent compound.

Table 2. Rate constants for isomerization of Dewer benzenes at 24,3 °C[1])

	$k \times 10^6$, sec^{-1}	ΔH, $\neq$ kcal	ΔS, $\neq$ eu
(structure)	5.18	23.0	− 5.0
(structure, Cl)	464	19.1	− 9.4
(structure, Cl...Cl)	0.0084	30.5	+12.0
(structure, Cl/F)	1860	—	—

1) Data from Breslow, Napierski and Schmidt [33].

In order to understand this phenonenon of partial forbiddeness, it is necessary to examine more closely the way in which orbital symmetry can creat a large activation energy. Activation energies, even for allowed reactions, exist primarily because of the Pauli exclusion principle. Each pair of electrons added to a collection of nuclei must occupy an orbital of successively higher energy. This usually means an orbital with one more nodal surface, or region where the wave function has a zero value.

When we speak of the symmetry of a wave function we mean simply the way in which the sign of the wave function changes from plus to minus as we go to different parts of the molecule. Hence symmetry is directly related to the nodal properties of the wave function.

A collection of nuclei will have a ground state wave function with a certain number of nodal surfaces of various kinds, σ and π. As the nuclei are rearranged to correspond to a chemical reaction, these nodal surfaces will be distorted and deformed. However, except for certain limiting cases, their number and kind will not change. In other words the wave function maintains its topological identity. This is the more generalized equivalent of conservation of orbital symmetry.

In the case of molecular orbitals, if a node exists in the region between two nuclei, this node will tend to persist even though the molecular orbital changes as a result of changing nuclear positions. Thus it is possible to follow a set of molecular orbitals during a concerted process by observing the constancy of their nodal patterns. This allows correlation diagrams to be drawn, even in the absence of symmetry (Zimmermann and Sousa [34]).

Furthermore, the chemical identity of the nuclei plays no direct part in determining the nodal patterns of the orbitals that they generate. The important feature is the number and kind of atomic orbitals that they contribute. Changing nuclei distorts and displaces the nodes, but they will still be between certain atoms in each orbital. Thus the MO's formed by two sets of different nuclei will be topologically equivalent if the same atomic orbitals are used.

A nodal surface raises the energy by a kinetic energy effect. However if the node occurs in a region where the orbital has a small value, the effect will be minimal. An example would be that of an anti-bonding MO for two atoms very far apart. As the atoms approach each other, the effect of a nodal surface between them becomes very great. In a chemical reaction, where atoms change their relative positions, some orbitals go up in energy and some go down because of the above phenomnenon. A reaction forbidden by orbital symmetry is simply one where electrons are trapped in orbitals that are going up in energy very rapidly because two or more atoms are approaching each other between which a node exists in the particular orbital.

Now it is clear that symmetry has only a minor effect on the above situation. Any similar collection of nuclei undergoing similar changes in relative inter-atomic distances will experience a similar increase in energy. Consider the reaction

$$
\begin{array}{ccc}
\text{R--C}{\equiv}\text{N} & \longrightarrow & \text{R--C}{=}\text{N} \\
+ & & \mid \quad \mid \\
\text{Cl--H} & & \text{Cl} \;\; \text{H} \\
& & \\
2\,a' & & 2\,a'
\end{array}
\tag{27}
$$

which contains, at most, a plane of symmetry. All bonds made or broken are symmetric, or a' with respect to this plane. Nevertheless we can see that the reaction is partly forbidden by symmetry. The nodal pattern of (27) is topologically the same as for the forbidden reactions (9) or (10). Thus we can often tell if a reaction has symmetry restraints upon it, even if lacking in symmetry, by relating it to a similar, but symmetric process.

As already noted, reduction in symmetry does reduce the magnitude of the energy barrier for a forbidden process. One way of explaining this is to realize that overlaps of orbitals which were zero in the symmetric case, become non-zero, but small, in the less symmetric case. This is largely due to the differences in electronegativity of different nuclei. For example, if all molecules were 100% ionic, consisting of positively or negatively charged spheres, all reactions would be allowed. Orbital symmetry restrictions would vanish, because they are a consequence of covalent bonding.

We have seen how perturbation theory describes a reaction as the mixing of orbitals, or the flow of electrons from filled MO's into empty MO's, in such a way as to correspond to the making and breaking of the desired bonds. A net positive overlap is necessary for this electron flow, or mixing of orbitals, to occur. The regions of positive overlap are regions of increased electron density, towards which nuclei will move. Atoms connected by a region of positive overlap will be bonded together.

Even in highly unsymmetrical molecules the above picture can be used to visualize the course of chemical reactions, and to draw conclusions about mechanism. It is necessary to have information about the nodal properties of the valence shell molecular orbitals. For many purposes, bond orbitals and anti-bond orbitals will suffice.

B. Some Reactions of Organometallic Compounds

Let us use the above ideas to make deductions about the detailed mechanisms of some important reactions of transition metal organometallic compounds. Most of these have molecules of very low symmetry and group theoretic rules are of little help. Nevertheless, the molecular orbital theory of bonding is well developed in these compounds (Cotton [35a], Schläfer and Gliemann [35b]).

1. The Ligand Migration Reaction

The first reaction we will discuss is the ligand migration reaction, or, as it is sometimes called, the insertion reaction. This well defined elementary reaction is a step in many important reactions of organometallics, including several catalytic processes of industrial importance. For a general review see Basolo and Pearson [36], or Schrauzer [37].

The following examples are written in steps to show the mechanisms as far as we know them.

$$CH_3Mn(CO)_5 \xrightarrow{\text{slow}} CH_3COMn(CO)_4 \qquad (28)$$
$$\text{18e} \qquad\qquad\qquad \text{16e}$$

$$CH_3COMn(CO)_4 + L \xrightarrow{\text{fast}} CH_3COMn(CO)_4L \qquad (29)$$
$$\text{16e} \qquad\qquad\qquad\qquad \text{18e}$$

(L is a phosphine, amine, arsine, CO, etc.)

$$HPt\ Cl(PEt_3)_2 + C_2H_4 \xrightarrow{\text{fast}} HPt\ Cl(PEt_3)_2C_2H_4 \qquad (30)$$
$$\underset{16e}{} \qquad\qquad\qquad \underset{18e}{}$$

$$HPt\ Cl(PEt_3)_2C_2H_4 \xrightarrow{\text{slow}} PtCl(PEt_3)_2C_2H_5 \qquad (31)$$
$$\underset{18e}{} \qquad\qquad\qquad \underset{16e}{}$$

The number of electrons in the valence shell of each complex or intermediate is shown. This illustrates the important rule that diamagnetic complexes of the transition metals from Groups IV—VIII have either sixteen or eighteen electrons in their outer shell. This rule is well known for stable complexes, being a modification of the inert gas rule. In addition Tolman [38] has recently pointed out that it works quite well for reactive intermediates or transition states. Exceptions are known, however.

Studies in which the molecule has a stereochemically informative group show that the main process is a motion of the CH_3 in (28), or the H in (31), towards the coordinated CO or C_2H_4 ligand. However some motion of the unsaturated ligand towards the anion facilitates the process (Glyde and Mawby [39]).

Fig. 12 shows what must be the critical orbitals. They are a filled bonding orbital connecting an anionic ligand, indicated as L, to the metal atom, and an empty anti-bonding orbital between the metal and the unsaturated ligand. This orbital is the anti-bonding partner to the π bonding orbital of the metal-olefin bond. In addition there is a σ type bond between the metal and olefin, shown only as a line.

Electron flow from the filled to the empty orbital as shown accomplishes the following: the metal-ligand bond is broken, the metal-olefin π bond is

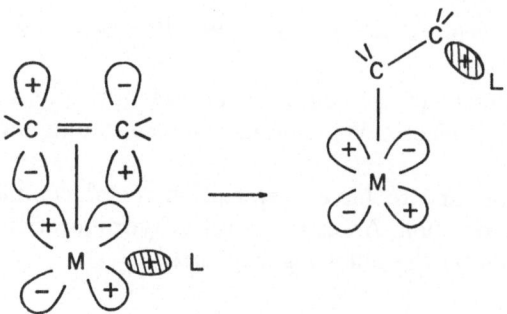

L migrates as anion

Fig. 12. Detailed MO representation of the mechanism of the ligand migration reaction. Electrons move from filled bonding orbital between M and L to empty anti-bonding orbital between M and olefin. Filled orbitals shaded

broken, the carbon-carbon double bond is broken, a bond is formed between L and the carbon atom of the olefin closest to it. A region of positive overlap causes L to move upward and the olefin to move to the right. The final product is as shown, a metal-alkyl bonded complex.

Trial and error shows that no other choice of orbitals will accomplish all of the necessary functions. For example, use of the filled metal-olefin π bond, and the empty M-L anti-bond orbital will strengthen the carbon-carbon double bond, instead of destroying it. The direction of electron motion shows clearly that L migrates with a pair of electrons, that is, as an anion. This means that the previously neutral olefin becomes anionic as it is converted to a coordinated alkyl group.

The mechanism shown in Fig. 11 is stereochemically detailed. It shows that the olefin and the migrating ligand L, must lie in the same plane, and in a *cis* arrangement to each other. The ligand L adds to the olefin on the side closest to it, in agreement with experiment. That is, the metal and L add *cis* to the olefin, as shown by Heck [40] and by Henry [41]. Also if L is an alkyl group containing an asymmetric carbon bonded to the metal, migration occurs with retention of configuration at that carbon (see Whitesides and Boschetto [42] and Hines and Stille [43]).

The latter information has actually only been obtained for migration of an alkyl group to a coordinated CO molecule. However the orbital analysis of this reaction is very much like that of Fig. 12, except that CO is coordinated in a linear fashion. If one attempts a similar analysis for coordinated SO_2, one discovers that it is impossible. SO_2 bonds to an electron rich transition metal by σ bonding only. There is no π bonding, and no empty anti-bonding orbital of π^* type. Interestingly enough, it is found that the reaction

$$R^*Fe(CO)_2C_5H_5 + SO_2 \longrightarrow R^*SO_2Fe(CO)_2C_5H_5 \qquad (32)$$

occurs with *inversion* of configuration at the optically active center R* (Whitesides and Boschetto [44]). The mechanism is unknown, but it is not that of Fig. 12.

It is not necessary to have a transition metal complex to have the ligand migration reaction. An example is the hydroboration reaction, which probably proceeds by the following sequence.

$$\qquad (33)$$

The key step is the movement of a pair of electrons from the B-H bonding orbital into the empty π^* orbital of the olefin. This is the mechanism postulated by Jones [45], and is completely analogous to the process of Fig. 12.

2. Oxidative-addition Reaction

A second general class of reaction of organometallic compounds of the transition metals is the oxidative-addition reaction. [36,37] In this reaction complexes of $(d)^8$ and $(d)^{10}$ configuration metal atoms add molecules XY in such a way that the metal is oxidized to $(d)^6$ or $(d)^8$ respectively. At the same time X and Y are added as ligands. Some examples would be

$$\text{Ir(CO)Cl[P(C}_6\text{H}_5)_3]_2 + \text{CH}_3\text{I} \longrightarrow \text{Ir(CO)Cl[P(C}_6\text{H}_5)_3]_2 \text{ CH}_3\text{I} \qquad (34)$$

$$\text{Pt(PR}_3)_3 + \text{HCl} \longrightarrow \text{PtHCl(PR}_3)_2 + \text{PR}_3 \qquad (35)$$

Thus the four-coordinated iridium(I) complex becomes a six-coordinated iridium(III) complex. The methyl ligand and the iodide ion add *trans* to each other (Muir and Pearson [46]).

It is obvious that the transition metal is acting as a source of electrons in these reactions, that is, as a reducing agent, or as a nucleophile. We start by considering the reactions of alkyl halides (or hydrogen halides. for that matter) with any nucleophile, B.

$$\text{B} + \text{CH}_3\text{Cl} \longrightarrow \text{BCH}_3^+ + \text{Cl}^- \qquad (36)$$

This class of reaction has been discussed by Kato, Morukama and Fukui, using perturbation theory. [48] The results are schematically illustrated in Fig. 13.

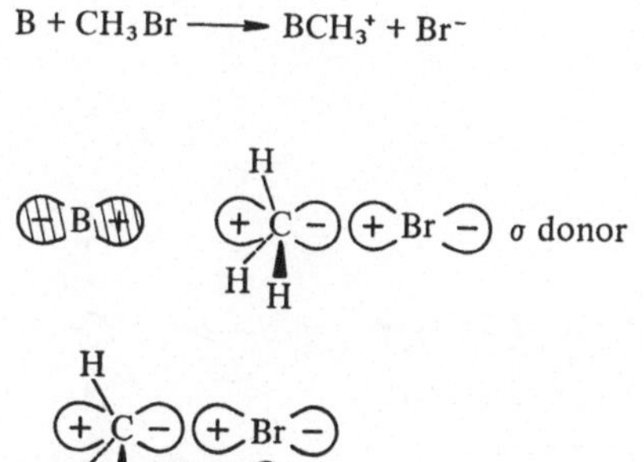

Fig. 13. Molecular orbital picture of classical SN2 mechanism for σ donor, and new mechanism for π donor. Filled orbitals are shaded

105

The critical LUMO is an anti-bonding σ^* orbital largely concentrated between carbon and chlorine. We assume the HOMO is in a p_σ orbital on the nucleophile, B. The usual SN2 displacement mechanism is predicted with inversion of configuration at carbon. Electrons flow from p_σ to σ^*, and in so doing break the C-Cl bond, and create a new B-C bond.

However, if we consider a π type donor atom as the nucleophile, we see a new possibility. Attack can occur at the front side of the alkyl halide, with retention of configuration at carbon. (Fig. 13) A d orbital on B would be best for this type of interaction, but a p orbital might also be effective.

For the π-type of interaction, these carbon-chlorine bond is broken again, but both CH_3 and Cl are bonded to B. This corresponds to what is observed in oxidative-addition. Fig. 14 shows in more detail the orbital symmetry aspects of addition reactions of XY to an iridium(I) complex.

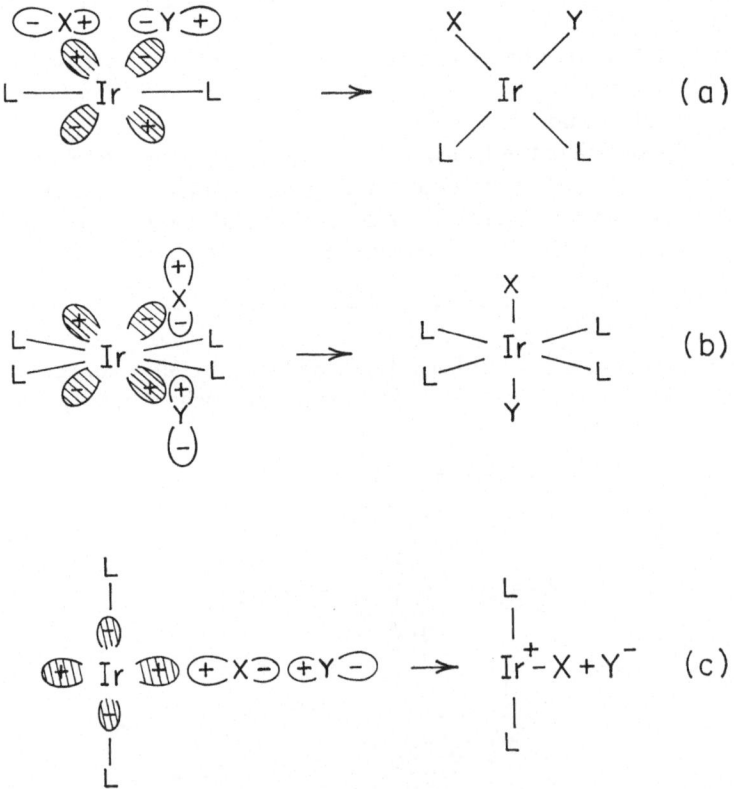

Fig. 14. Three possible mechanisms for oxidative-addition reaction allowed by orbital symmetry: (a) *cis* addition with retention at X, (b) *trans* addition with retention at X, (c) two step mechanism with inversion at X

There is considerable evidence that some oxidative-addition reactions are one-step, concerted processes. [46] Both *cis* and *trans* addition of XY, are found, as predicted in Fig. 14 [b]. Also retention of configuration at X has been found in a number of cases. These are cases where X is an optically active alkyl or silyl group and Y is hydrogen or another alkyl group (Sommer, Lyons and Fujimoto [47a], Walborsky and Allen [47b], James and Ng [47c], Whitesides *et al.* [47d]).

In the case where XY is an alkyl halide, another mechanism is possible. This is the normal nucleophilic substitution with inversion at carbon (see Fig. 14 c), followed by the addition of the halide ion in a second step. There is some evidence against this mechanism, but the critical test of determination of configuration at carbon remains undone. When Y is a poor leaving group such as H^- or CH_3^-, of course the two-step mechanism is impossible, and concerted addition is the only plausible mechanism.

Oxidative-addition reactions of alkyl halides also occur with metal ions such as thallium(I). In this case, the HOMO is an *s* orbital on the metal of σ symmetry. The concerted mechanism is now impossible and only a two-step reaction can occur. The same is true for reactions of metals in the free state, such as zinc or magnesium, with alkyl halides. Note that free radicals may be formed, instead of ions, in all cases discussed. Free radical reactions are usually allowed by symmetry.

3. Oxidative-cyclization Reaction

Earlier (Section V) we discussed how two olefin molecules coordinated to a metal atom could undergo disproportionation. There is another process that could occur, provided the metal atom had a sufficiently high reduction potential. This is oxidative-cyclization, first observed by Collman *et al.* [48] and also by Ashley Smith, Green and Stone. [49] Acetylenes can be cyclized as well as olefins.

Assuming that coordination is a necessary first step, some examples would be

$$IrCl(PR_3)_2R(C\equiv CR)_2 \longrightarrow \begin{array}{c} R \ \ R \\ | \ \ | \\ PR_3 \ \ C=C \\ | \diagup \ \ \ | \\ Cl-Ir \ \ \ \ \ \ \\ | \diagdown \ \ \ | \\ PR_3 \ \ C=C \\ | \ \ | \\ R \ \ R \end{array} \qquad (37)$$

18e   16e

[b] The bond symmetry rule, however, shows that concerted *trans* addition is forbidden. It is also unlikely for steric reasons.

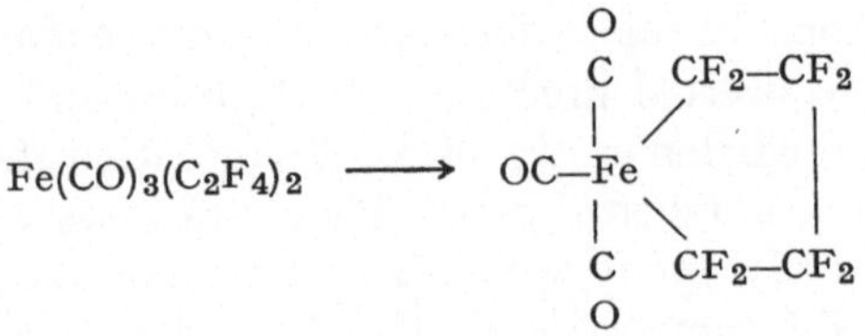

$$\text{Fe(CO)}_3(\text{C}_2\text{F}_4)_2 \quad \longrightarrow \quad \underset{\text{16e}}{\overset{\displaystyle \overset{\text{O}}{\underset{\displaystyle \parallel}{\text{C}}} \quad \text{CF}_2\!-\!\text{CF}_2}{\underset{\displaystyle \overset{\text{C}}{\underset{\text{O}}{\parallel}}\quad \text{CF}_2\!-\!\text{CF}_2}{\text{OC}\!-\!\text{Fe}}}} \qquad (38)$$

$$\underset{\text{18e}}{}$$

Notice that the metal atom is oxidized by two units in each case. The two electrons go into the unsaturated ligands and help produce a carbon-carbon σ bond. At the same time the π bonding of the olefins, or acetylenes, to the metal is destroyed, as are the π bonds themselves. Two σ bonds remain between the metal and two carbon atoms.

Just as in the olefin disproportionation reaction (Fig. 10), it is necessary to put electron density into the b_2^* orbital. This destroys the carbon-carbon π bonding, the metal-ligand π bonding, and creates the new carbon-carbon σ bond. However, instead of taking electrons from the b_1 orbital of the complex, let us now take it from a filled d orbital concentrated on the metal of a_1 symmetry.

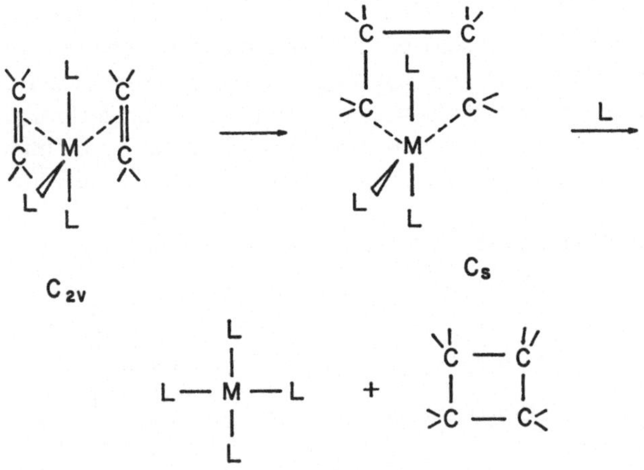

$$A_1 \times B_2 = B_2$$

Fig. 15. Steps in the cyclization of two olefin molecules to a cyclobutane molecule, catalyzed by a metal complex. First step is oxidative-cyclization, which occurs by B_2 reaction coordinate

Fig. 15 shows the result of such an electron transfer, or mixing of orbitals of a_1 and b_2 symmetry. The reaction coordinate is predicted to be of B_2 symmetry. According to Fig. 7, this corresponds to an up or down displacement out of the plane. That is, the two top carbon atoms of the olefin molecules would move away from the metal, and the two bottom carbon atoms would move towards the metal. The result would be the metallocycle structure shown in Fig. 15, of C_S symmetry.

Also shown in the figure is a possible further reaction in the presence of excess ligand, L. The reverse of an oxidative-addition reaction can occur. This process, called reductive-elimination, leads to the elimination of a cyclobutane molecule, plus the formation of a metal complex in which the metal atom has been reduced back to its original oxidation state.

This sequence has been shown because there are examples known of substituted olefins being cyclized to cyclobutane derivatives by certain transition metal complexes (Schrauzer [50a], Heimbach, Jolly and Wilke[50b]). In addition, metal complexes will catalyze the opening of cyclobutane rings to olefins and other rearranged products, as shown particularly by Cassar, Eaton and Halpern [51]. It is likely that the sequence shown in Fig. 15, both forward and reverse, is the mechanism for many of these reactions. Metal ions such as silver(I) which do not undergo oxidation readily probably involve a different mechanism (Paquette [52a], Gassmann [52b]).

VII. Photochemical Reactions

The subject of symmetry rules for reactions of excited electronic states is a large and complicated one. At this time only a few general comments will be made. The theoretical Eqs. (1)—(5) are still valid, but E_n and ψ_n must interchange with E_0 and ψ_0, where n refers to the excited state. We note that processes which occur readily in the ground state may become difficult in the excited state. This is because the term (E_0-E_n), which is negative, becomes replaced by (E_n-E_0), which is positive, and raises the energy in Eq. (4).

Conversely processes which are difficult, or forbidden, in the ground state may be facile, or allowed, in the excited state. One change will most certainly occur. The nuclei will rearrange somewhat to give a new molecular geometry for the excited state. These structural changes can be predicted with some success by perturbation theory (Pearson [53a], Devaquet [53b]).

Lack of detailed knowledge of excited states of most molecules makes the understanding of photochemical reactions difficult. However, definite predictions can be made, both by perturbation theory and by correlation methods, if we are allowed to select which excited state is involved.

109

For example, we can easily tell how to make a forbidden reaction in the ground state, allowed photochemically. Consider a simple reaction, hydrogen deuterium exchange

$$H_2 + D_2 \longrightarrow 2\,HD \tag{39}$$

Like other reactions of diatomic molecules this reaction is forbidden by symmetry to occur by a four-center transition state. In terms of the bond symmetry rules

$$
\begin{array}{ccccc}
\text{H—H} & & \text{H--H} & & \text{H}\quad\text{H} \\
+ & \longrightarrow & \vdots\;\;\vdots & \longrightarrow & |\;+\;| \\
\text{D—D} & & \text{D--D} & & \text{D}\quad\text{D} \\[4pt]
a_1 + b_2 & & & & a_1 + b_1
\end{array}
\tag{40}
$$

a b_2 orbital must become a b_1 orbital.

Now we imagine that we excite one molecule, H_2 or D_2, into the state $(\sigma_g)(\sigma_u{}^*) = \Sigma_u^+$, by putting one electron into the anti-bonding σ_u orbital. On approaching an unexcited molecule, D_2 or H_2, the configuration is no longer given by $(a_1)^2(b_2)^2$ as in (40). Instead we have $(a_1)^2(b_2)(b_1)$, since the linear combinations of two σ_u^* orbitals give a_2 and b_1 symmetry, of which the latter has lower energy.

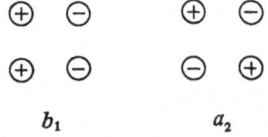

$$b_1 \qquad\qquad a_2$$

Now we allow reaction to occur to give 2 HD molecules. We form products also in the $(a_1)^2(b_1)(b_2)$ configuration, one molecule of HD being in the excited Σ state. The symmetries of all the orbitals containing electrons are matched up, and the exchange is allowed with a four center transition state.

The thermal decomposition of phosgene occurs by a complex free radical route, just as does the reverse reaction (15). It can easily be seen that $(b_1) \rightarrow (a_1)$ excitation is necessary to make the forbidden direct decomposition into CO and Cl_2 occur. This does not correspond to the lowest excited state of phosgene, which is a B_2 state due to $n - \pi^*$ excitation. Instead reaction would require an n to σ^* excitation. Experimentally $COCl_2$ and $COBr_2$ do decompose photochemically and apparently by a molecular process. [22] However the excited state involved is not known.

Acknowledgement. Sincere thanks are given to the National Science Foundation for support of this work under grant GP-31060 X.

VIII. References

[1] Woodward, R. B., Hoffmann, R.: The conservation of orbital symmetry. Weinheim: Verlag Chemie 1970.
[2] Mulliken, R. S.: Rev. Mod. Phys. *4*, 1 (1932).
[3] Shuler, K. E.: J. Chem. Phys. *21*, 624 (1953).
[4] a) Griffing, V.: J. Chem. Phys. *23*, 1015 (1955); b) Vanderslice, J. T.: J. Chem. Phys. *23*, 1035, 1039 (1955).
[5] Coulson, C. A., Longuet-Higgins, H. C.: Proc. Roy. Soc. (London) *A 191*, 39; *A 192*, 16 (1947).
[6] Fukui, K., Yonezawa, T., Shingu, H.: J. Chem. Phys. *20*, 722 (1952); Topics Curr. Chem. *15*, 1 (1970).
[7] Dewar, M. J. S.: J. Am. Chem. Soc. *74*, 3341 *et seq.* (1952).
[8] Bader, R. F. W.: Can. J. Chem. *40*, 1164 (1962).
[9] Salem, L., Wright, J. S.: J. Am. Chem. Soc. *91*, 5947 (1969).
[10] a) Pearson, R. G.: J. Am. Chem. Soc. *91*, 1252, 4947 (1969); b) Pearson, R. G.: J. Chem. Phys. *52*, 2167 (1970).
[11] a) Pearson, R. G.: Theoret. Chim. Acta *16*, 107 (1970); b) Pearson, R. G.: Acc. Chem. Res., *4*, 152 (1971); c) Pearson, R. G.: Pure Appl. Chem. *27*, 145 (1971).
[12] a) Fukui, K.: Bull. Chem. Soc. Japan *39*, 498 (1966); b) Fukui, K.: Accounts Chem. Res. *4*, 57 (1971); c) Klopman, G., Hudson, R. F.: Theoret. Chim. Acta *7*, 165 (1967); d) Salem, L.: Am. Chem. Soc. *90*, 543 553 (1968).
[13] Salem, L.: Chem. Phys. Letters *3*, 99 (1969).
[14] England, W., Salmon, L. S., Ruedenberg, K.: Topics Curr. Chem. Fortschr. Chem. Forsch. *23*, 31 (1971).
[15] Thompson, H. B.: Inorg. Chem. *7*, 604 (1968).
[16] Gillespie, R. J.: J. Chem. Educ. *40*, 295 (1963).; J. Chem. Soc. *1963*, 4672.
[17] Pearson, R. G.: J. Am. Chem. Soc. *94*, 8287 (1972).
[18] Wigner, E., Witmer, E.: Z. Physik *51*, 859 (1928).
[19] Fletcher, E. A., Dahneke, B. E.: J. Am. Chem. Soc. *91*. 1603 (1969).
[20] a) Hildebrand, J. H.: J. Am. Chem. Soc. *68*, 915 (1946); b) Schweitzer, P., Noyes, R. M.: J. Am. Chem. Soc. *93*, 3561 (1971).
[21] Minn, F. L., Hanratty, A. B.: Theoret. Chim. Acta *19*, 390 (1971).
[22] Bamford, C. H., Tipper, C. F. H.: Comprehensive chemical kinetics. Vol. 4. Amsterdam: Elsevier Publishing Co. 1972.
[23] Mahan, B. H.: J. Chem. Phys. *55*, 1436 (1971).
[24] Uemura, S., Sasaki, O., Okana, M.: Chem. Commun. *1971*, 1064.
[25] Hoffmann, R., Howell, J. M., Muetterties, E. L.: J. Am. Chem. Soc. *94*, 3047 (1971).
[26] Dewar, M. J. S.: The molecular orbital theory of organic chemistry, p. 340. New York: McGraw-Hill Book Co. 1969.
[27] Archer, E. M., van Schalkwyk, T. G. D.: Acta Cryst. *6*, 88 (1953).
[28] Littler, J. S.: Tetrahedron *27*, 81 (1971).
[29] Herzberg, G.: Molecular spectra and molecular structure, Vol. III, p. 585. Princeton: D. Van Nostrand Co., Inc. 1967.
[30] a) Hughes, W. B.: J. Am. Chem. Soc. *92*, 532 (1970); b) Mango, F. D., Schachtschneider, J. H.: J. Am. Chem. Soc. *93*, 1123 (1971); c) Lewandos, G. S., Pettit, R.: J. Am. Chem. Soc. *93*, 7087 (1971).
[31] Caldow, G. L., MacGregor, R. A.: J. Chem. Soc. A *1971*, 1654.
[32] Pignolet, L. H., Horrocks, W. D., Holm, R. H.: J. Am. Chem. Soc. *92*, 1855 (1970).
[33] Breslow, R., Napierski, J., Schmidt, A. H.: J. Anm. Chem. Soc. *1972*, 5906.
[34] Zimmermann, H. E., Sousa, L. R.: J. Am. Chem. Soc. *94*, 834 (1972).

35) a) Cotton, F. A.: Chemical applications of group theory (Second Editor). New York: John Wiley and Sons, Inc. 1971; b) Schläfer, H. L., Gliemann, G.: Einführung in die Ligandenfeldtheorie. Frankfurt/M.: Akademische Verlagsgesellschaft 1967.

36) Basolo, F., Pearson, R. G.: Mechanisms of inorganic reactions. New York: John Wiley and Sons, Inc. 1967.

37) Schrauzer, G. N.: Transition metals in homogeneous catalysis. New York: Marcel Dekker, Inc. 1971.

38) Tolman, C. A.: Chem. Soc. Rev. *1*, 337 (1972).

39) Glyde, R. W., Mawby, R. J.: Inorg. Chim. Acta *5*, 317 (1971).

40) a) Heck, R. F.: J. Am. Chem. Soc. *91*, 6707 (1969); b) Heck, R. F.: J. Am. Chem. Soc. *90*, 5518, 5535 (1968).

41) Henry, P. M.: J. Org. Chem. *37*, 2443 (1972).

42) Whitesides, G. M., Boschetto, D. J.: J. Am. Chem. Soc. *91*, 4313 (1969).

43) Hines, L. F., Stille, J. K.: J. Am. Chem. Soc. *92*, 485, 1798 (1972).

44) Whitesides, G. M., Boschetto, D. J.: J. Am. Chem. Soc. *93*, 1529 (1971).

45) Jones, P. R.: J. Org. Chem. *37*, 1886 (1972).

46) Pearson, R. G., Muir, W. F.: J. Am. Chem. Soc. *92*, 5519 (1970).

47) a) Sommer, L. H., Lyons, J. E., Fujimoto, H.: J. Am. Chem. Soc. *91*, 7051 (1969); b) Walborsky, H. M., Allen, L. E.: Tetrahedron Letters *11*, 823 (1970); c) James, B. R., Ng, F. T. T.: J. Chem. Soc. A *1972*, 355; d) Whitesides, G. M., San Filippo, J., Stedronsky, E. R., Casey, C. P.: J. Am. Chem. Soc. *91*, 6542 (1969).

48) Collman, J. P., Kang, J. W., Little, W. F., Sullivan, M. F.: Inorg. Chem. *7*, 1298 (1968).

49) Ashley Smith, J., Green, M., Stone, F. G. A.: J. Chem. Soc. A *1969*, 3019.

50) a) Schrauzer, G. N.: Advan. Catalysis *18*, 373 (1968); b) Heimbach, P., Jolly, P. W., Wilke, G.: Adv. Organometal. Chem. *8*, 29 (1970).

51) a) Cassar, L., Eaton, P. A., Halpern, J.: J. Am. Chem. Soc. *92*, 3515 (1970); b) Cassar, L., Eaton, P. A., Halpern, J.: *92*, 6366 (1970); c) Byrd, J. E., Cassar, L., Eaton, P. A., Halpern, J.: Chem. Commun. *1971*, 40.

52) a) Paquette, L. A.: Accounts Chem. Res. *4*, 280 (1971); b) Gassman, P. G., Atkins, T. J.: J. Am. Chem. Soc. *93*, 4597 (1971).

53) a) Pearson, R. G.: Chem. Phys. Letters *10*, 31 (1971); b) Devaquet, A.: J. Am. Chem. Soc. *94*, 5626 (1972).

Received February 27, 1973

The Discovery of Organic Synthetic Routes by Computer

Prof. Herbert Gelernter, Dr. Natesa S. Sridharan*, Arthur J. Hart, and Shou-Chian Yen

Department of Computer Science, S.U.N.Y. at Stony Brook, New York, USA

Prof. Frank W. Fowler and Ho-Jane Shue**

Department of Chemistry, S.U.N.Y. at Stony Brook, New York, USA

Contents

---

* Present address: Department of Computer Science, Stanford University, California, USA.

** Present address: Brookhaven Nat. Lab., Upton, New York, USA.

A computer program written for the IBM System 360/67 has successfully discovered multistep syntheses for relatively complex organic structures without on-line guidance or intercession on the part of the chemist-user. The program is able to deal with a wide variety of functional and structural features. Information concerning organic synthetic reaction mechanisms is provided to the computer in a tabular form reaction library containing, for each reaction, structural schema for the target and subgoal molecules, and a set of tests, largely heuristic, to govern the choice of reaction. With its initial quite limited library of reactions and problem-solving heuristics, the program developed a conceptually correct synthesis for the complex polycyclic twistane ring system.

I. Introduction

It is the purpose of this article to report the present status of a research program in progress at Stony Brook University which has as its goal the design and development of a system of computer programs which, in concert, will be able to discover novel and useful synthesis routes for organic structures of interest to chemists. Because we view the problem as an immensely challenging exercise in artificial intelligence, we wish our system (henceforth, SYNCHEM, for SYNthetic CHEMistry) to develop its results without the direct intervention of the user-chemist, although user intervention should certainly be provided for in a production system. For reasons which we shall discuss below, such a free-running system offers advantages from the chemist's point of view as well.

At this time, some four years after the inception of our project, we report that SYNCHEM has discovered multi-step syntheses for several dozens of relatively complex target structures. The proposed synthetic routes range in quality from probably good and valid to embarassingly naive, with most falling in the center of this scale. One synthesis in particular, for the polycyclic bridged molecule of twistanone, is of special interest, and has been selected to bear the burden of illustrating how our program works. Although the synthesis procedures for twistanone were discovered by the grace of a modicum of serendipity and programmer intervention (the latter to compensate for the lack of some problem-solving tree search heuristics that have yet to be incorporated into the system), they provide evidence that our approach to the problem is a reasonable one that is capable of dealing with organic structures of current interest to the scientific community, and that our ultimate goal of a system of practical value to organic chemists is in fact feasible.

The general outlines for this research program were first enunciated by the senior author at the International Symposium on Artificial Intelligence held at Sukhanovo, U.S.S.R. in February, 1967. At that time, the conviction was expressed that while the theoretical underpinnings of artificial intelligence as a scientific discipline left much to be desired, the techniques and methodology of artificial intelligence in general, and heuristic programming in particular, had matured to the point where problems of substantial content and of interest in themselves rather than mere vehicles for artificial intelligence research ought to be selected by those engaged in that activity. It was suggested that workers in artificial intelligence turn their attention to the physical sciences to find a rich breeding ground for problems upon which they might ply their trade to the mutual benefit of both their own discipline and the external field upon which their work might impinge. The example given was the problem of organic synthesis discovery.

115

The design of synthetic routes to complex organic molecules is a good problem for the study of artificial intelligence from several points of view. First, no knowledgable chemist would deny that the discovery of good organic synthesis procedures requires a high order of inventiveness, insight, inductive and deductive subproblem manipulation, hypothesis generation and evaluation, and in fact, just about every activity identified with intelligent problem-solving behavior. But the problem has other attractive characteristics. It builds upon existing foundations in the state of the art of artificial intelligence and leads into presently unexplored areas of inquiry. The discovery and identification of good and useful organic synthesis routes requires a combination of both analytic and synthetic problem-solving procedures. In the terminology introduced by Amarel [1], it is a combined problem of derivation and formation. Such problems are important because many "real life" problems are of this type. As a class, this category is only beginning to receive the attention it merits, largely because of its more complex structure. Just as earlier work on problems of derivation has made the solution generation, or *analytic search* part of our problem relatively straightforward to manage, we would expect that progress in the solution evaluation, or *synthetic reconstruction* phase of our program will be of value to others undertaking research in areas where problems of formation play a significant but non-independent role.

It is worth pointing out, too, that our approach to the problem is open-ended in that once an initial measure of success in finding syntheses for simple structures has been attained, we expect that the program's domain of competence may be extended by extending the then-current working program. By enlarging the set of reaction mechanisms, schema, and heuristics embodied in the program, we enlarge the set of functional groups and structural features the program may encompass, or the set of alternative procedures for synthesizing a given feature. In effect, the difference between a feasibility demonstration for the application of artificial intelligence to organic synthesis discovery and a system of genuine utility to chemists is one of degree rather than of kind.

Ours is by no means the only artificial intelligence research group to seek vehicles for our investigations among the natural sciences. Indeed, one of the earliest and most fruitful projects of that genre, the Heuristic Dendral program for the determination of molecular structure from NMR and mass spectra being developed at Stanford [2], antedates our own activity by several years. The particular problem of organic synthesis discovery by computer, both with and without user intervention, has independently attracted the attention of several research groups whose primary interests lie in organic chemistry per se, rather than in artificial intelligence [3-5] and substantial and impressive progress in the computer-assisted discovery of synthetic routes has been reported in the literature [6]. It will be seen that our point of view, which is fundamentally different from that of the chemist-directed groups, has caused our work to evolve along somewhat different lines than theirs, although it is clear that our overall conceptions of the nature of the problem are remarkably similar.

The results reported here mark the attainment of the first phase of a research program whose ultimate objective is a system which will take as input some representation of the target molecule to be synthesized together with a list of conditions and constraints that must govern the solution of the problem. The output is to be, in general, a set of proposed synthesis procedures for the input structures, each of which starts with available compounds and reagents taken from a "shelf library" specified by the input conditions, and each of which is annotated with estimated yields for each step of the procedure together with byproduct predictions and target molecule separation procedures (with estimated separation efficiency) for each step. The system is assumed to have at its disposal a reaction library providing generalized procedures for the synthesis of functional groups and structural features, both singly and in multiple configurations (we call such synthesis-relevant entities *synthemes*), and a complete shelf library listing every compound considered to be available as a starting material. Each item in the shelf library carries a list of conditions on its presumed availability (*i.e.*, cost, handling hazards, availability in industrial, pilot, or laboratory quantities, on-hand availability, and so on). Every synthesis proposed by the system (several possible routes will generally be developed for a given target molecule) is to have an overall ad hoc estimate of the degree to which the procedure satisfies the conditions and constraints of the problem. An effective problem-solving system will, of course, eliminate from consideration those pathways for which the estimated merit is significantly lower than the best computed at a given point in the search for the best possible synthesis.

It will be seen that the computing effort of the problem-solving system is divided (not necessarily equally) between the tasks of solution generation and solution evaluation. Allowing for the great difference in the natures of the problems considered, solution generation does not differ vastly for organic synthesis from the procedures described by Gelernter for theorem proving in Euclidean geometry [7]. In broad outline, the processes coincide: selection of the subgoal for development (logical expressions in geometry, organic molecules in SYNCHEM), selection of subgoal-generating schema (theorems in geometry, reaction mechanisms in SYNCHEM), generation of the next level of subgoals, checking for terminating conditions (subgoals coinciding with premises in geometry, compounds listed in the shelf library in SYNCHEM), and pruning the problem-solving tree of unpromising and redundant subgoals. The major departures in procedure occur in selection of subgoal-generating schema and in pruning the problem-solving tree.

117

Because our computer-compatible formalization of organic chemistry is so much more complicated than that which sufficed for Euclidean geometry, the cost (in terms of computing effort) of wanton subgoal generation is substantially greater for SYNCHEM than for geometry, so that investment in the contraception of unwanted and invalid subgoals brings correspondingly greater returns for SYNCHEM. The difficulty in controlling the growth of SYNCHEM's problem-solving tree is further exacerbated by the fact that criteria for pruning valid subgoals from the tree in order to suppress the development or circular, redundant, or otherwise unsatisfactory pathways are considerably less well defined for organic chemistry than was the case for geometry.

Bringing our synthesis-discovery program to the stage where it could develop reasonable problem-solving trees containing a high density of complete and satisfactory syntheses within the limits of an incomplete reaction library was designated as the first benchmark for our research effort, and we present below evidence that we have reached that stage. After describing our initial results in solution generation, we will indicate what remains to be done and the problems yet to be solved, particularly those bearing on the solution evaluation phase of our program, which we have barely broached.

II. The Synthesis-Discovery Algorithm

A companion paper [8] describes the subgoal-generating and synthesis-discovery subsystems of SYNCHEM in sufficient detail to enable the interested reader to make a critical evaluation of this work. We briefly synopsize these programs here to clarify the results to follow.

Most often, input to SYNCHEM is in the form of the Wiswesser line-formula representation of the target molecule, although a kind of connection-matrix representation of the molecule (our so-called "Topological-Structural Description", or TSD) is also accepted by the program. Internally, SYNCHEM manipulates the entities of organic chemistry in both Wiswesser and TSD representations, using the most appropriate form for the purpose at hand, and converting freely between the two. The TSD is most often used for structure analysis, manipulation, and transformation. The Wiswesser line-formula notation (WLN), which provides a canonical name for any molecular structure, is ideal for information storage and retrieval, comparing newly-generated subgoals with those already on the problem-solving tree and on the list of available compounds (the shelf library), and I/0 processes.

After input, the target compound TSD is analysed for synthesis-relevant functional groups and structural features (synthemes), and some of these are selected for development. Corresponding to each syntheme is a chapter of the reaction library, each chapter comprising an arbitrary number of reaction schema for the synthesis of that particular syntheme, either by itself, or in combination with other synthemes in frequently-occurring and well-understood configurations. A particular syntheme having been selected, the appropriate chapter of the reaction library is brought into the computer. Each schema of the chapter is provided with a set of tests to be performed on the target molecule. These tests embody many of the chemistry heuristics that guide the program. Based on the results of these tests, the program may reject the given schema, modify the ad hoc merit rating for that reaction (for example, the reaction merit might be raised if a conjugated activating group is present, lowered if steric hindrance is detected), the reaction procedure may be modified (a different reagent might be specified in the presence of groups sensitive to the usual reagent), or protection procedures may be initiated for sensitive groups.

Programmed by the adjusted set of reaction schema selected for the target molecule, SYNCHEM generates a set of subgoals for that molecule. For each of these, an ad hoc overall merit is computed based on the adjusted reaction merits and an estimate of the complexity of the subgoal molecules. If a synthesis-search terminating condition has not been signalled, the "best" available subgoal is selected for further development, and the procedure is recursively continued. A synthesis-search tree (SYNCHEM's problem-solving tree) is thereby generated, the structure of which depends upon the currently active algorithms for subgoal merit computation, and for determining which subgoal shall be called the best for further consideration. A synthesis has been completed when a path has been generated linking the target molecule with the catalog of available compounds.

The initial reaction library contains, in varying stages of completeness, chapters for the synthesis of aldehydes, ketones, alcohols, organic acids, esters, halides, acid halides, Grignard and other organo-metallic reagents, nitriles, the olefin bond, and six-carbon ring formation. Multi-functional syntheses are indexed under each of the synthemes dealt with by that reaction.

Our shelf library is the catalog of organic compounds and reagents available from the Aldrich Chemical Supply Company, augmented by a number of commonly obtainable materials that Aldrich doesn't list (acetone, for example). The Aldrich catalog was selected simply because it was available on punched cards in WLN representation. Unfortunately, the cards provide only the catalog number along with the WLN names in machine readable form, so that the arduous task of adding the information we require concerning cost, physical properties, quantities available, and so on for the more than 8,000 items in the Aldrich catalog alone still remains to be done before our initial list of available compounds meets our specifications for SYN-CHEM's shelf library.

The system is programmed in PL/1 augmented by an extensive library of specially-written subroutines which greatly facilitate the manipulation of the list-like data structures SYNCHEM uses to represent organic molecules, problem-solving trees, reaction schema, and much of the program's working storage. PL/1 was selected as our programming language because it provides a number of features designed to make list-processing relatively convenient — pointer operations, based and offset variables, area allocation, and so on. Four years of programming experience has given us no reason to regret our choice of language. The program runs at Stony Brook on an IBM 360/67, requiring a minimum partition size of 360 K for a production synthesis-search in its present version. All data concerning computation times were obtained in this minimum partition, where SYNCHEM is forced to perform extensive program overlay and data structure swapping operations.

Twistanone (Tricyclo[4.4.0.0 ^{3,8}]decan-2-one)

Although SYNCHEM's performance with the target molecule twistanone was sullied by a certain amount of good fortune, and on-line intervention was necessary to prevent the program from wasting inordinate amounts of time on barren branches of the problem-solving tree (but only after SYN-CHEM had already discovered its first two proposed syntheses on its own), our results for twistanone are the most interesting we have achieved to date. The good fortune to which we refer was a consequence of the fact that our TSD–WLN interconversion routines (which are massive and intricate symbol-manipulation programs) comprise almost all, but not quite the full set of Wiswesser canons, which requires a volume of several hundred pages for its complete explication [9]. When SYNCHEM cannot perform consistent WLN-TSD transformations on a given structure, the offending molecule is dropped from the search tree. This unintentional pruning occurred only at the first level of the twistanone synthesis tree (where extremely complex subgoals were generated), but it turned out to be quite efficacious in removing several undesireable subgoals while preserving the satisfactory ones. When we discuss the complete synthesis-search tree below, we shall point out where intervention was necessary to get SYNCHEM back on a fruitful track. It is of course in the nature of exploratory research in artificial intelligence that one discovers and formulates improved tree pruning and search algorithms by contemplating the failings of those in current use. As is the case in training the human intellect, this is a continuous and never-ending process.

One final point needs to be made before we get on with our results. A further contribution to limiting the excessive growth of the search tree may be traced to the limited number of schema in our initial reaction library. Clearly the more ways the program is able to deal with a given syntheme, the larger the potential set of subgoals for a given molecule. On the other hand, because the set of chemistry heuristics associated with each reaction schema is also incomplete, many unsatisfactory subgoals were accepted on the search tree that will be rejected by later versions of SYN-CHEM provided with more discriminatory heuristics in its reaction library. We live and learn.

Referring now to Fig. 1, the string heading the printout, L666/BI/EJ A A 2BF J AVTJ, is the WLN representation for twistanone, and is a direct transcription of the input card for this problem. This is followed by a severely edited output of the TSD connection matrix to which the machine converted the input WLN. A complete SYNCHEM TSD contains

```
TSD FOR STRING REAC IN

L666/BI/EJ A A 2BF J AVTJ

ICT CELL CCDE BND1:M BND2:M BND3:M BND4:M BND5:M BND6:M
  GO    8   V   1C:1   18:1
  GO   1C   C    8:1   12:1   24:1    H:1
  GO   12   C   1C:1   14:1    H:1    H:1
  GO   14   C   12:1   16:1    H:1    H:1
  GO   16   C   14:1   18:1   26:1    H:1
  GO   18   C   16:1   20:1    8:1    H:1
  GO   20   C   18:1   22:1    H:1    H:1
  GO   22   C   20:1   24:1    H:1    H:1
  GO   24   C   22:1   26:1   1C:1    H:1
  GO   26   C   24:1   16:1    H:1    H:1
```

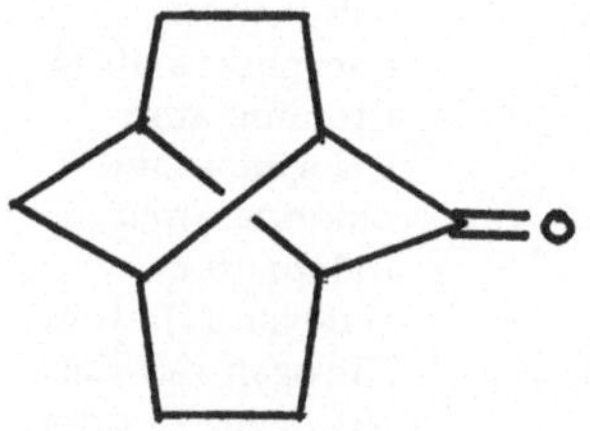

```
------------------------------
|  ATTRIBUTE TABLE LISTING   |
------------------------------

        ATTRIBUTE     CELL#

    CARBCCYCLICRING    8   18   2C   22   24   1C

    CARBOCYCLICRING    8   18   2C   22   24   26   16   14   12   10

    CARBCCYCLICRING    8   18   16   26   24   10

    CARBCCYCLICRINC   18   16   26   24   22   2C

    CARBCCYCLICRING   16   14   12   1C   24   26

    CARBCCYCLICRING   18   16   14   12   10   24   22   2C

    CARBCCYCLICRING    8   18   16   14   12   1C

    CARBCNYL  CRCUP    8

         KETCNE        8
```

$$\text{tricyclo}\left[4.4.0.0^{3,8}\right]\text{decan-2-one}$$

(Twistanone)

L666/BI/EJ A A 2BF J AVTJ FG=8.CCCC CAFECNS=0.C0OC MERIT=3.8750

Fig. 1. Topological-structural description (TSD) and attribute table listing for twist-anone

a good deal more information than is displayed in the edited output, including an extensive attribute table, which also appears in abbreviated form following the connection matrix in Fig. 1. A detailed description of SYN-CHEM's internal TSD is given in Ref. [8]. The attribute table following the connection matrix in the edited output lists each one of the seven cyclic paths in the twistanone molecule, the node location of the carbonyl functionality, and the location of the ketone syntheme. The carbonyl functionality (which, in general, might be a component of a number of different synthemes) is listed independent of its possible associated synthemes to provide information for various conversion routines. The complete set of features listed in the primary attribute table for a given molecule is displayed in Table 1. When a particular TSD has been selected for further develop-

Table 1. Primary attribute table property list. Capital letters in parentheses give the Wiswesser designation for that property

1	Oxygen atom	21	Olefin bond
2	Hydroxy group	22	Variable ($n)
3	Carbonyl group	23	Alkyl chain
4	Dioxo (W) group	24	Carbocyclic ring
5	Nitrogen (N) atom	25	Heterocyclic ring
6	Sulfur atom	26	Epoxide
7	Phosphorus atom	27	Carboxylic acid
8	Bromine atom	28	Aldehyde
9	Fluorine atom	29	Alcohol
10	Chlorine atom	30	Ether
11	Iodine atom	31	Anhydride
12	Halogen (J) atom	32	Bicyclic ring system
13	Nitrogen (M) group	33	Amide
14	Nitrogen (Z) group	34	Ketene
15	Nitrogen (K) group	35	Acid halide
16	Phenyl ring	36	Halide
17	Hetero atom	37	Ketone
18	Acetylenic bond	38	Ester
19	Other triple bond	39	Amine
20	Other double bond	40	Nitrile

ment, (becoming a *generating subgoal*), additional descriptive information for that molecule is extracted from the TSD in response to particular "questions" asked by the heuristic tests associated with the reaction schema selected for that subgoal. This secondary information is appended to the primary table, and may comprise any synthetically relevant features whatever, either following some standard characterization, or defined for a particular heuristic test for a particular reaction schema on the spur of the moment. Examples of such secondary attributes in current use are location and susceptibility of acid, base, and water sensitive groups, nodes that tend to be electron withdrawing or electron donating, molecular symmetries, and so on.

Following the attribute table in Fig. 1 is the WLN for the generating subgoal as reconverted from the TSD. This string must of course match the input WLN (or, in general, the WLN selected from the problem-solving tree as the generating subgoal). It was the failure of this consistency test for a few extremely complicated structures that produced the fortuitous tree pruning mentioned above.

Printed on the same line as the reconverted WLN, the following expressions appear: FG $=8$, CARBONS $=0$, MERIT $=3.875$. The quantity MERIT as it appears in this context is intended to reflect an ad hoc estimate of the complexity of the molecule, on a scale where 10 indicates utter simplicity, and zero indicates maximum complexity. (In the printout of the synthesis search tree, this quantity is labeled CPMERIT.) MERIT is an ad hoc function of FG, which is intended to be a count of the number of distinct complications in the structure, and CARBONS, the number of non-cyclic carbons in the molecule (cyclic carbons are reflected in the quantity FG). At the moment, FG is defined to be the number of synthemes listed on the primary attribute table uniformly weighted (in this case, the seven carbocyclic rings and the ketone group). Since there are no non-cyclic carbons in twistanone, CARBONS $=0$. For the syntheses described in this paper, MERIT was computed with the function:

$$\text{MERIT} = \text{CPMERIT} = \tfrac{10}{1.2}\,[\exp.(-0.1385 * \text{FG}) + 0.2\exp(-0.0346 * \text{CARBONS})]$$

These definitions have been and will continue to be freely modified as we gain experience. Since MERIT is intended to provide an estimate of the anticipated difficulty of synthesis for the given subgoal molecule, it is arbitrarily set to 10 for compounds discovered in the shelf library.

Having completed its preliminary processing of the target (or subsequently, the selected subgoal) compound, the program chooses a syntheme from the attribute table and calls in the appropriate chapter of the reaction library for developing that syntheme. Fig. 2 is the beginning segment of the output trace of SYNCHEM's activity in developing the ketone group in twistanone. The 40-bit string labeled ABITS OF GOAL synopsizes the attribute table of the generating subgoal. In this case, the ones in positions 3, 24, and 37 signal the presence of at least one carbonyl group, carbocyclic ring, and ketone group, respectively. Each chapter of the reaction library is headed by a table of contents listing a 40-bit string, the ABITS FOR SCHEMA (n) for each active reaction schema in the chapter. These index strings enable the program to immediately discard those reactions which will always fail for generating subgoals with specific attributes. Referring

```
CURRENT CHAPTER IS KETCNE

ABITS OF GOAL        001C0C0CCC0CCCCCCCCC0GGC1CC00000C0C001C00
ATEST FCR.SCHEMA  1  CGGCGCGCGCCCCCCCCCCCCGCC00001CC0CCCC00
ATEST FOR SCHEMA  2  COCOOCGCCCOGCCCCCCCCCCCCCC000CCC0C0CC000
ATEST FOR SCHEMA  5  CCCOCCOCCCCCCCCCCCCCCCCOC11111CCC00CC101
ATEST FOR SCHEMA  6  CCCOCCCCOGCCCCCCCCCGCCCCCCG0CCCC0CCCC00
ATEST FCR SCHEMA  7  CCCCCCCCCCCCCCCGOCCCCCCCOCG01CCCOC0OC000
ATEST FOR SCHEMA  8  CCCOC0O1011CCCCC01C01OC001C011CC0C0CC010
ATEST FOR SCHEMA  9  COCOC0O1C11CCCCCOCCCCCCGC01111CC0C0C0O1
ATEST FCR SCHEMA 10  0CCGCCCCCCCCCCCCCCCCCCCCC000000000CC000
ATEST FCR SCHEMA 11  CCCCCCCCCCCCCCCCCCGCCCCCCCCCCCCCCC00CC00

ATEST SUCCEEDED

ELIGIBLE_SCHEMA      '110C11111111C0CC0'B



CURRENT SCHEMA NUMBER IS                              1

GCAL PATTERN  $1.V$3.

SUBGOAL PATTERN  $1YC$3

KETCNE LIBRARY  ** CXIDATION USING CHRCMIC CXIDE ** DESC1        SEXIT   2     FEXIT   2
EASE  10  NSTEPS   1


ALCO1              NCT SUCCESSFUL/ 2
FOR 41 RESULT = COCCCCCCC0COC0CCCCCCCCCCCCCCCOCO


ALCO2              NOT SUCCESSFUL/ 2


CYCLIC ALKANE      NCT SUCCESSFUL/ 2
TEST NOT CEFINEC


CLEFIN    BCNC    USE JONES REAGENT AND RESTRICT TC SMALL SCALE
                                        -2     0     0


OXIDE             USE PYRICINE AS SCLVENT
                                      0     0     0


ALCEHYDE          PROTECT THRU ACETAL AND USE PYRIDNE AS SCLVENT
                                       -5     2     0


OICL              PFCTECT THRU. CYCLIC ETHER AND USE JONES REAGENT
                                       -5     2     C


CYCLIC KETCNE     PROTECT THRU CYCLIC ETHER AND USE JONES REAGENT
                                       -5     2     0
BRANCHING TC CYCLICKETCNE
TEST SUBRCUTINE NCT PFCGRAMMEC
FOR 45 RESULT = CCCCCCCCCCC0CCCCCCCCCCCCCCCC0CO


ANCTHER KETCNE    PROTECT THRU CYCLIC ETHER AND USE JCNES REAGENT
                                       -5     2     0
BRANCHING TC ANCTHER KETCNE
FCR 44 RESULT = COCCCOC0O0OCCCCCCCCCCCCCCCC0CCCO


CARBOXYLIC ACID   PFCTECT THRU ESTER          -5     2     0

CHOOSING CELL  8 CF GOAL FOR MATCHING.  1 MATCHES FOLND.
L666/CH B B 2AC JTJ BC

SUBGOAL GENERATEC L666/CH B B 2AC JTJ BC

L666/CH B E 2AC JTJ BQ  NOT AVAILABLE
```

Fig. 2. Beginning of the output trace for the first page of the synthesis search tree for twistanone

124

to Fig. 2, the one in position 30 of ABITS FOR SCHEMA 1 indicates that reaction 1 (oxidation of an alcohol to the ketone) will always fail in the presence of an ether group. Schema 5 (Grignard-nitrile reaction) with ABITS in positions 26 through 30, 38, and 40, will fail as well in the presence of the epoxide, carboxylic acid, aldehyde, alcohol, ester, or another nitrile. In this particular instance, no reaction was discarded as a consequence of these so-called A-tests. This is of course not generally the case. Fig. 7, extracted from a later stage of the output trace, illustrates the more usual circumstance where some reactions are in fact eliminated from consideration beforehand by the A-tests. But here, the ELIGIBLE SCHEMA bit string indicates that every active schema in the chapter represents a potential subgoal-generating reaction.

The trace next records SYNCHEM's results with each of the eligible schema, starting, in this case, with Schema 1 for the synthesis of a ketone group by oxidation of an alcohol. The interested reader is again referred to Ref. [8] for a detailed explanation of the goal and subgoal patterns, the heuristic testing and subgoal generating procedures, and the numerous TSD-manipulation subroutines implied by the above. Here it will suffice to explain briefly the more important and interesting items in the record.

In the goal and subgoal patterns, the symbols N, where N is an integer, stand for molecular structure variables. In the literature of organic chemistry, this symbol is usually written R_N. When N is even, N may be matched to any substructure whatever; when N is odd, hydrogen matches for N are excluded. The patterns are printed out in WLN representation, with V standing for a carbonyl group, Y a single-branched alkyl carbon, and Q a hydroxy group. In the chemistry literature, they would be written thus:

$$\text{Goal Pattern } R_1-\overset{\overset{\displaystyle O}{\|}}{C}-R_3 \qquad \text{Subgoal Pattern } R_1-\overset{\overset{\displaystyle OH}{|}}{C}H-R_3 .$$

The periods following each of the structure variables in the WLN goal pattern indicate to the matching routines that in this case (because of symmetry in both goal and subgoal patterns), the reverse match ($3 and $1 matched to the structures originally matched to $1 and $3, respectively) will produce an identical subgoal, and need not be executed.

Then, following the name of the schema (chromic oxide is mentioned as the primary oxidizing agent), the notation EASE 10, NSTEPS 1 appears. The quantity EASE is intended to reflect the chemist's ad hoc estimate of the overall desireability of including a step in the synthesis procedure derived through that schema. The starting value for EASE (on a scale where 10 is excellent, zero is hopeless) takes into consideration probable yield, convenience, reliability, and any personal bias the chemist preparing the entry for the reaction library might have for or against that procedure. The quantity NSTEPS is just the number of distinct chemical processes (possibly requiring separation, purification, or concentration operations after each step) to effect the transformation of that schema when no modifications in procedure are required to deal with interfering functionalities. Both these quantities are subject to alteration to reflect the results of a sequence of tests (called N-tests) that are performed on the generating subgoal before a match is attempted. After the match has been executed, the generated subgoals may be subjected to a second sequence of tests (called F-tests) which may dictate further alterations in EASE or NSTEPS. A possible consequence of the N-tests and F-tests is, of course, the outright rejection of the schema or of a particular set of generated subgoals.

In our reaction library as it is presently constituted, the starting values of EASE for all of the schema, as well as the modifications in EASE dictated by the test results, are little more than informed guesses. We are not especially confident that they now reflect the relative desireability of the different schema in a chapter for synthesizing a given syntheme; we have even less reason to believe that the values are consistent from chapter to chapter. A great deal of chemist's work remains to be done after the mechanics of solution generation have been mastered. This is perhaps a good time to reiterate our earlier caveat — our current reaction library is not only incomplete in terms of the synthemes for which chapters have yet to be included, but several of the existing chapters have extremely lean rosters of schema, and many of the schema have incomplete sequences of N-tests and F-tests.

Returning to Fig. 2, the output trace continues with a listing of N-tests applied to the generating subgoal. In this case, no test result was positive, and so EASE and NSTEPS remained at their starting values, and the match of the generating subgoal to goal pattern was executed (as it so happened here, successfully). While we do not go into detail in describing the N-test record (indeed, the volatility of the reaction library would superannuate such detail long before publication), the following points are of interest. The first two tests check for the presence of a primary or secondary alcohol group in the generating subgoal. The notation NOT SUCCESSFUL/2 indicates that if a positive result is obtained (*i.e.*, the alcohol group is detected in the generating subgoal), this reaction will fail to produce useful subgoals, all activity with this schema is to be terminated, and the schema specified after the slash (in this case, Schema 2) is to be selected next. Since it cannot be determined from the attribute table whether an alcohol is primary, secondary, or tertiary, an N-test subroutine must be called to extract the information from the TSD. Subroutine 41 is the appropriate one in this case; it adds the information concerning the degree of each alcohol group to the attribute table extension, and returns a bit string abstracting the result. The trace output line FOR 41 RESULT $= 0$ indicates to the program that neither primary, secondary, nor tertiary alcohol groups were detected.

The next N-test checks for secondary alcohols, with no result (hence negative) indicated. Note that a call for subroutine 41 was not necessary here, since the desired information was discovered on the extended attribute table, where it had been placed in an earlier step. Since tertiary alcohols cannot be oxidized to the ketone, they do not affect this reaction, and no test is made for them. We remark that it is because this reaction will go in the presence of tertiary alcohols or diols (which can be protected with a cyclic ether, as specified by the seventh N-test in the list) that a simple A-test will not serve to exclude generating subgoals with primary or secondary alcohols.

The DIOL test is an example of an EASE or NSTEPS modifying procedure. If a suitable configuration of two alcohol groups is detected, the instructions PROTECT THRU CYCLIC ETHER AND USE JONES REAGENT are appended at the appropriate place in any completed syntheses passing through this branch of the search tree, EASE is decremented by 5, and NSTEPS is incremented by 2, as specified by the values following the procedure modification instructions listed in the trace. NSTEPS must of course be increased to allow for the additional operations of protecting the sensitive group and later removal of the protecting functionality. On the other hand, if the OLEFIN BOND test produces a positive result (note that no subroutine call appears if the test can be performed on the attribute table), the instructions USE JONES REAGENT AND RESTRICT TO SMALL SCALE are appended, EASE is decreased by 2, but NSTEPS remains unchanged, since no additional operations are necessary.

Following the record of N-tests, the trace indicates that SYNCHEM attempted a match of the goal pattern to the generating subgoal, aligning the ketone group in the pattern with the ketone group in cell 8 of the generating subgoal TSD, and that the match succeeded (fourth line from the bottom of the output). The generated subgoal WLN follows (L 666/CH B B 2AC JTJ BQ), together with the remark that the material was not listed as available on the shelf library.

A similar sequence appears in the trace output for each active schema in the chapter, and the entire procedure is duplicated for each chapter selected for the generating subgoal. When a generated subgoal is accepted for inclusion on the problem-solving tree, the quantity MERIT, the ad hoc compound complexity, is computed for each compound comprising the subgoal (which, in general, might specify the reaction of several compounds). An ad hoc function of MERIT and EASE is used to estimate the overall subgoal merit (SGMERIT) of a given path through the search tree. Currently, the subgoal evaluation function is arbitrarily defined to be the product of the individual compound MERIT values and the reaction EASE (after adjustment), normalized to a maximum (best case) value of 10. For some runs, SGMERIT is diminished by a small quantity which is an increasing function of depth in the search tree to force SYNCHEM to prefer shorter synthesis routes. We have no reason to believe that our current subgoal evaluation function even remotely approaches that which would produce an optimum search regimen.

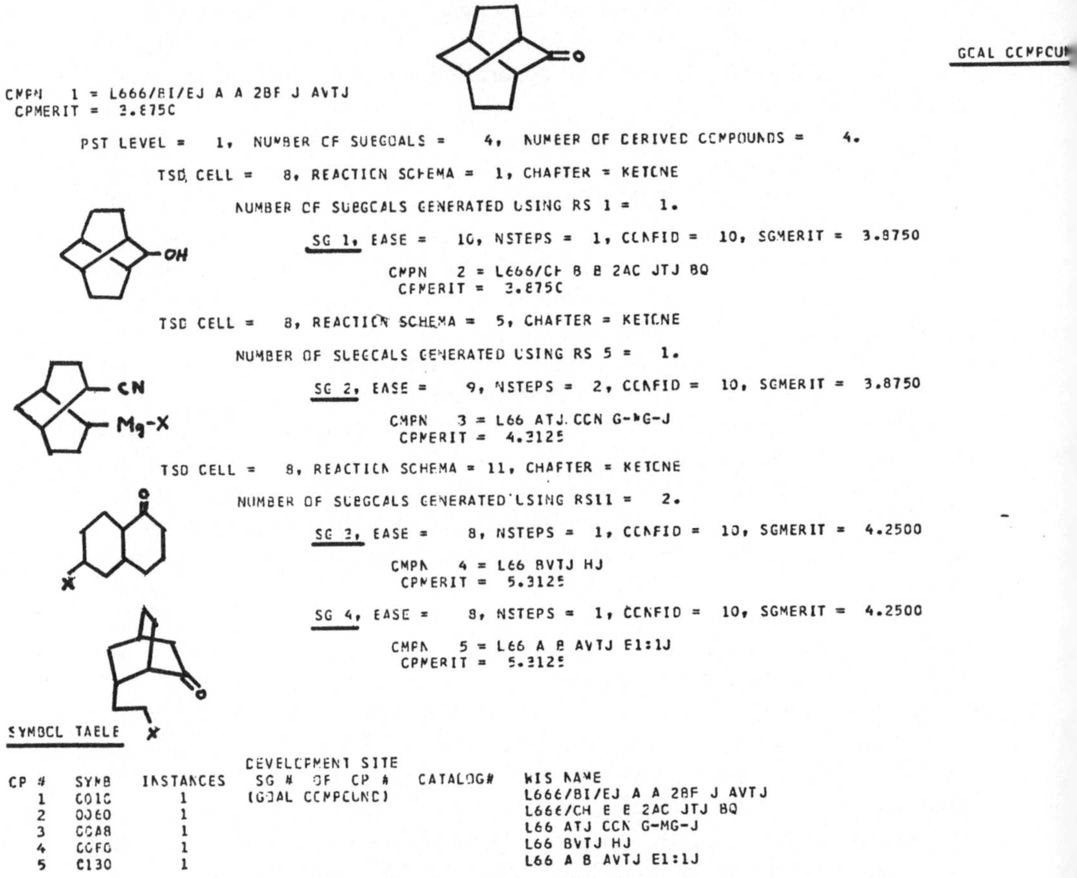

```
                                                                                      GCAL CCMPCU

CMPN   1 = L666/BI/EJ A A 2BF J AVTJ
 CPMERIT =  3.E75C

     PST LEVEL =   1,  NUMBER CF SUEGOALS =    4,  NUMEER OF CERIVEC CCMPOUNDS =     4.

         TSD CELL =    8,  REACTICN SCFEMA =  1,  CHAFTER = KETCNE

             NUMBER CF SUEGCALS CENERATEO USING RS 1 =    1.

                 SG 1,  EASE =   1C,  NSTEPS =  1,  CCNFID =   10,  SGMERIT =   3.8750

                     CMPN   2 = L666/CH B B 2AC JTJ BQ
                      CFMERIT =   3.E75C

         TSD CELL =    8,  REACTICN SCHEMA =  5,  CHAFTER = KETCNE

             NUMBER OF SLEGCALS CENERATEO USING RS 5 =    1.

                 SG 2,  EASE =    9,  NSTEPS =  2,  CCNFID =   10,  SGMERIT =   3.8750

                     CMPN   3 = L66 ATJ.CCN G-MG-J
                      CPMERIT =   4.3125

         TSD CELL =    8,  REACTICN SCHEMA =  11,  CHAFTER = KETCNE

             NUMBER OF SLEGCALS CENERATED USING RS11 =    2.

                 SG 3,  EASE =    8,  NSTEPS =  1,  CCNFID =   10,  SGMERIT =   4.2500

                     CMPN   4 = L66 BVTJ HJ
                      CPMERIT =   5.3125

                 SG 4,  EASE =    8,  NSTEPS =  1,  CCNFID =   10,  SGMERIT =   4.2500

                     CMPN   5 = L66 A B AVTJ E1:1J
                      CPMERIT =   5.3125

SYMBCL TAELE

CP #    SYMB    INSTANCES    SG # OF CP #    CATALOG#    HIS NAME
                            (GOAL CCMPCUNC)
  1     C01C       1                                     L666/BI/EJ A A 2BF J AVTJ
  2     0J60       1                                     L666/CH B B 2AC JTJ BQ
  3     CCA8       1                                     L66 ATJ CCN G-MG-J
  4     CCFC       1                                     L66 BVTJ HJ
  5     C130       1                                     L66 A B AVTJ E1:1J
```

Fig. 3. Symbol table and first page of the twistanone synthesis search tree

Fig. 3 is the synthesis-search tree output at the end of the first subgoal-generating cycle. The symbol table, which provides an index to the tree, lists five compounds at this stage, the target molecule and the four first-level subgoals which were accepted for further development. The internal symbol table includes, in addition to the development site information and Aldrich catalog number (for compounds discovered to be available in the shelf library) which are printed out, pointers for each compound indicating the location of each instance of that compound on the search tree. The information stored in the symbol table vastly facilitates traversal of the tree, data retrieval, and reconstruction of search routes through the tree.

With the following explanations, the printout of the first "page" of the tree itself is not difficult to comprehend. The item PST LEVEL indicates the problem-solving tree level of the page. TSD CELL designates the particular syntheme of the generating subgoal developed by the schema. In the ketone chapter, Reaction Schema 1 (RS=1) calls for the oxidation of an alcohol, Schema 5 is a Grignard-nitrile reaction, and Schema 11 is a ketone alkylation. We have already pointed out that the tree designation of the quantity referred to earlier as MERIT is CPMERIT. The item labeled CONFID is not presently used, and has arbitrarily been set to 10. It has been provided to enable chemists, at some later date, to express their confidence in the predicted results for a given reaction schema based on published and personal experience with a given reaction. This quantity, too, is subject to modification by N- and F-test. Later versions of SYNCHEM will make use of CONFID as a factor in the procedure for selecting the most likely routes from among the several which will normally be discovered by the synthesis-search programs. We do not intend to include CONFID in the computation for SGMERIT, for to do so would tend to exclude the bolder and more interesting, if riskier, explorations from the search tree.

Any chemist examining page 1 of the twistanone search tree will remark that Compound 3 of Subgoal 2 is a most unlikely Grignard reagent. Indeed, such Grignard reagents should be (and in due course, will be) excluded from the tree by an F-test. Since the N-tests for the Grignard generating schema usually exclude the offending subgoal at the next level, we have placed a rather low priority on the correction of this particular defect.

Fig. 4 is the output trace of the search through the tree for completed syntheses, and failing these, the search for the best subgoal for further development. The choice was easy to make here. Since at this stage the tree is of only one level, SYNCHEM arbitrarily selected one of the two generated subgoals which had the maximum value of SGMERIT, Compound 5, to become the new generating subgoal. The TSD for Compound 5 is printed out next, followed by its attribute table. Note that the customary chemist's notation "X" is used to designate an arbitrary halogen in the TSD, while the WLN and attribute table use the Wiswesser symbol "J" for that purpose. The next page of the search tree, Fig. 5, exhibits the 8 generated subgoals accepted for generating subgoal Compound 5 (L66 A B AVTJ E2J). The first three of these derive from the Halide chapter of the reaction

H. Gelernter *et al.*

```
*****   TREE SCAN RETURNEC TO ROCT


*** RETHREADING TRACE ***
LNSOLVED
UNSOLVED
UNSOLVED
UNSOLVED

*** END RETFRED ***
*** SCLMAX TRACE ***
MERITSCL=   C.CCCO
*** NSCLMAX TRACE ***   TEMP    000101C8
   4.2500   TEMP   CC01C178
   4.25CO   TEMP   CC01C110
   3.875C   TEMP   CC0100A8
   3.875C



   MERITNSCL=   4.250C
   NO ACCPETABLE SYNTFESIS FOUNC FCR CCMPCUND
   SEARCFING
   TEMP 00C101C8
   TEMP C0010178
   TEMP 00C10110
   TEMP CCC100A8
   TEMP FFC0CCCC
   BO NUDE 000101C8
   LAST BEST SC FFCCOOCO
   ADCRESS OF CRTGCAL= CCO1C1C8ADDRESS CF CRTCMP = 0001C1FO
   CRTCMP =     5(    5)
   SEARCHING
   TEMP FFCOCOCO
   BC NUDE FFCCCOOC
   LAST BEST SC FFCOCO00
   CURRENT CCMPCUND IS    5
```

Fig. 4. Output trace of the "best next subgoal" search on the first page of the twist-anone tree

library, the next five from the Ketone chapter. Halide reactions 1,2, and 3 represented on this page of the tree synthesize the halide from the alcohol, carboxylic acid, and olefin bond, respectively. Here again subgoals appear which more thorough N-testing will exclude.

Selection of the best subgoal for development is more difficult now, for SYNCHEM must decide whether to continue the branch of the tree on which it is currently active to a greater depth, or whether it would be more productive to backtrack to a different branch on a higher (or, for that matter, the same or lower) level. The tree exploration algorithm is among the more volatile ingredients in our current system; indeed, problem-solving tree exploration represents an entire subdiscipline of artificial intelligence. For the record, we describe briefly the algorithm that governed SYNCHEM in finding (or, in the instance we shall specify below, failing to find) the syntheses discussed in this article.

130

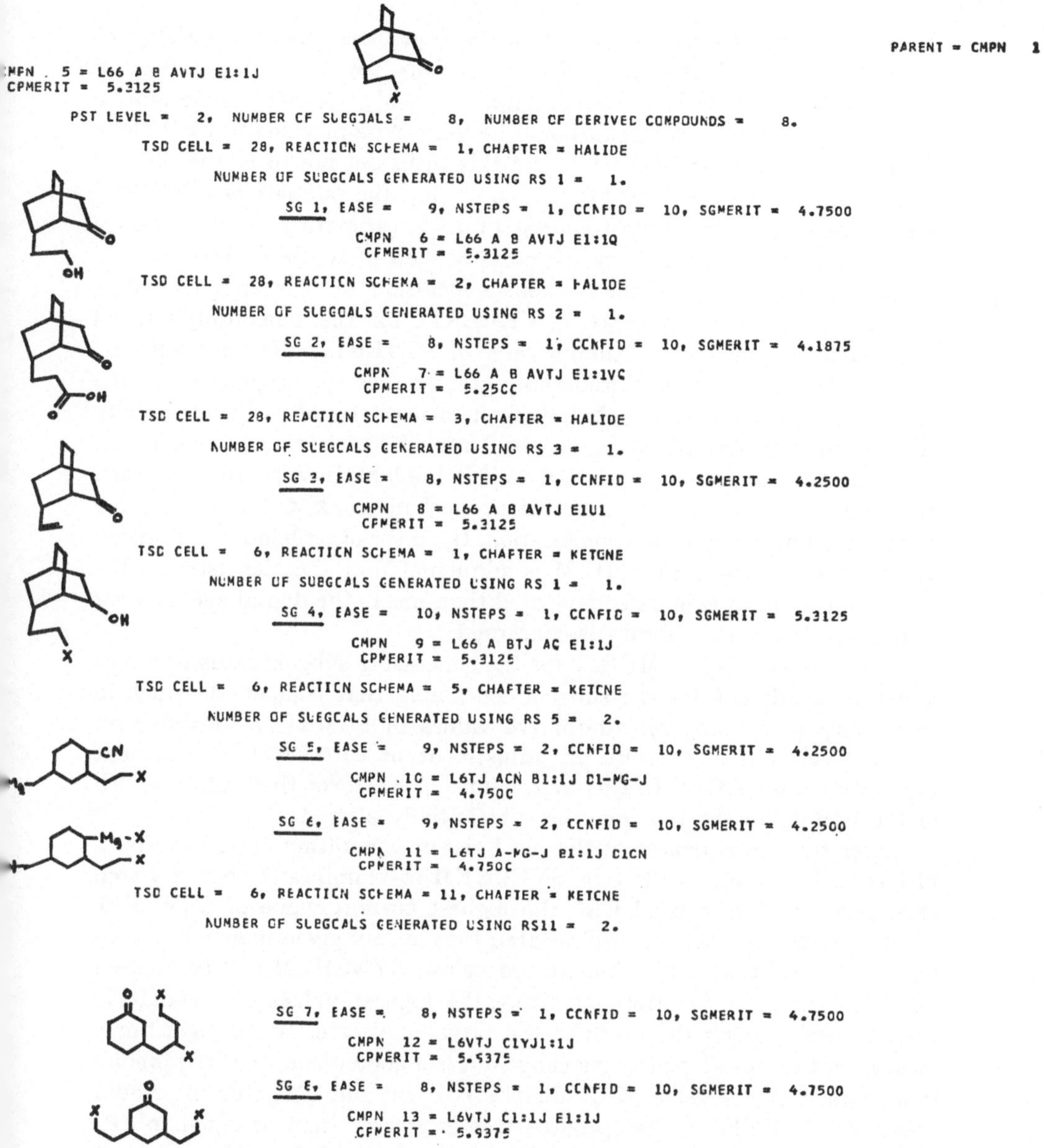

CMPN   5 = L66 A B AVTJ E1:1J
CPMERIT = 5.3125

PST LEVEL =   2,   NUMBER CF SLEGJALS =   8,   NUMBER CF CERIVEC COMPOUNDS =   8.

TSD CELL =   28,   REACTICN SCFEMA =   1, CHAPTER = HALIDE

NUMBER UF SUBGCALS GENERATED USING RS 1 =   1.

SG 1, EASE =   9, NSTEPS =   1, CCNFID =   10, SGMERIT =   4.7500

CMPN   6 = L66 A B AVTJ E1:1Q
CPMERIT =   5.3125

TSD CELL =   28,   REACTICN SCFEMA =   2, CHAPTER = HALIDE

NUMBER CF SLEGCALS GENERATED USING RS 2 =   1.

SG 2, EASE =   8, NSTEPS =   1, CCNFID =   10, SGMERIT =   4.1875

CMPN   7 = L66 A B AVTJ E1:1VC
CPMERIT =   5.25CC

TSD CELL =   28,   REACTICN SCFEMA =   3, CHAPTER = HALIDE

NUMBER CF SLEGCALS GENERATED USING RS 3 =   1.

SG 3, EASE =   8, NSTEPS =   1, CCNFID =   10, SGMERIT =   4.2500

CMPN   8 = L66 A B AVTJ E1U1
CPMERIT =   5.3125

TSD CELL =   6,   REACTICN SCFEMA =   1, CHAPTER = KETCNE

NUMBER UF SUBGCALS GENERATED USING RS 1 =   1.

SG 4, EASE =   10, NSTEPS =   1, CCNFID =   10, SGMERIT =   5.3125

CMPN   9 = L66 A BTJ AC E1:1J
CPMERIT =   5.3125

TSD CELL =   6,   REACTICN SCFEMA =   5, CHAPTER = KETCNE

NUMBER OF SLEGCALS GENERATED USING RS 5 =   2.

SG 5, EASE =   9, NSTEPS =   2, CCNFID =   10, SGMERIT =   4.2500

CMPN 1C = L6TJ ACN B1:1J C1-MG-J
CPMERIT =   4.750C

SG 6, EASE =   9, NSTEPS =   2, CCNFID =   10, SGMERIT =   4.2500

CMPN   11 = L6TJ A-MG-J B1:1J C1CN
CPMERIT =   4.750C

TSD CELL =   6,   REACTICN SCFEMA = 11, CHAPTER = KETCNE

NUMBER CF SLEGCALS GENERATED USING RS11 =   2.

SG 7, EASE =   8, NSTEPS =   1, CCNFID =   10, SGMERIT =   4.7500

CMPN   12 = L6VTJ C1YJ1:1J
CPMERIT =   5.9375

SG E, EASE =   8, NSTEPS =   1, CCNFID =   10, SGMERIT =   4.7500

CMPN   13 = L6VTJ C1:1J E1:1J
CPMERIT = · 5.9375

Fig. 5. Second page of the twistanone synthesis search tree

A basic ad hoc heuristic tenet guiding the exploration of the tree derives from our assumption that CPMERIT, a measure of the complexity of a given organic molecule on the tree, is a good predictor of the difficulty to be encountered in finding a suitable synthesis route for that molecule.

If this were indeed the case, then in the best of all possible worlds, for suitably correlated and normalized functions for the computation of CPMERIT and SGMERIT, the maximum value of SGMERIT in the subtree sprouted for a given molecule should be very close in value to CPMERIT for that generating subgoal. If in fact this turns out not to be the case for a given subtree, it is reasonable to assume that the estimate of CPMERIT was incorrect, either overestimating or underestimating the complexity of the generating subgoal from the point of view of synthesis discovery.

SYNCHEM executes the obvious conclusion of this analysis. After each cycle of subgoal generation, CPMERIT for the generating subgoal is adjusted to equal the maximum value of SGMERIT in the new subtree. In effect, if that value is considerably greater that the original value, the algorithm is acknowledging that the generating subgoal was synthetically simpler than it seemed to be on the basis of its structural complexity, probably because something similar to it is available as a starting material, or because a highly specific and efficient mechanism exists for synthesizing a required multi-functional configuration. If, on the other hand, the adjusted value is much lower, SYNCHEM is admitting that the chemistry of the generating subgoal is more complicated than meets the digital eye, at least insofar as its reaction library is concerned.

The new value of CPMERIT for the generating subgoal leads to a new value of SGMERIT for the subgoal containing that compound, which in turn, may (depending on alternative values of SGMERIT at this next higher level subtree) require an adjustment of CPMERIT at the next higher level above that. In this way, figures of merit for that entire branch of the tree up to the root may be substantially altered.

After the consequences of the most recent generating cycle have been propagated throughout the tree, SYNCHEM re-examines the entire search tree, selecting the subgoal with the highest current value of SGMERIT for further development. It will be seen that at any given moment, except for the proviso that will be mentioned below, SYNCHEM will be focused on that branch of the tree exhibiting the highest values of SGMERIT, level by level, along the route to the target molecule. If the algorithm's most recent choice of generating subgoal was a good one in that the prediction implicit in its value of CPMERIT is borne out (*i.e.*, the maximum value of SGMERIT in the sprouted tree is greater than or equal to CP-MERIT), the new subtree will contain the subgoal of choice for the next cycle, and exploration will continue down that branch. (Recall however, that our present function for computing SGMERIT penalizes subgoals for excessive depth in the tree, so that depth-first search cannot continue indefinitely, even if that mode isn't excluded by the merit-adjustment algorithm.) It is generally the case that a branch is continued when reaction schema produce subgoals that split the generating subgoal into two or more smaller substructures (with consequent higher values of CPMERIT), and *a fortiori* so when one of the substructures is discovered on the list of available compounds, giving it a value of 10 for CPMERIT.

If, on the other hand, the most recent choice of generating subgoal was a poor one, necessitating downward revision of CPMERIT for the generating compound (and hence, downward revision of SGMERIT for the generating subgoal), SYNCHEM may refocus its attention anywhere on the search tree, at any level, on any branch. Here we introduce the proviso mentioned above. To prevent the program from wildly prancing about the tree in response to very mild degradation in SGMERIT in the branch being explored, a programmer-adjustable threshold (presently set at 0.5) in SGMERIT decay must be exceeded before SYNCHEM will shift its attention to a different branch of the tree. In this way, minor lapses in vision on the part of our figure-of-merit evaluation functions are excused.

A few final remarks complete our brief description of the search algorithm. Through the cross-referencing index of the symbol table, adjusted values for CPMERIT are transmitted to every instance of the affected compound in the tree. Although this procedure can result in a higher value of SGMERIT appearing in a remote branch of the tree than that for the generating subgoal in focus, it is not permitted to sidetrack SYNCHEM from its current branch when things are going well (*i.e.*, values of SGMERIT along the route are monotonically increasing). Pathways to complete syntheses must of course be suitably labeled, to prevent the program from continually rediscovering them.

We now turn our attention to the terminating synthesis routes discovered by SYNCHEM after a total of approximately twenty minutes of search time (Fig. 6). The twistane ring system was first prepared by Whitlock [10] in 1962 and other syntheses [11] have since been published. Conceptually, the computer discovered all of these approaches, but due to presently severe limitations on our reaction library, they differ in details. SYNCHEM's syntheses need to be further refined by a knowledgeable chemist before they could be considered useful.

The last step in these published synthetic routes (*i.e.*, the first set of subgoals in the problem-solving tree, which is, of course, a record of "working backwards") is the intramolecular alkylation of a ketone[a]. Although

---

[a] Reaction Schema 11 for Ketone Alkylation.

$$-\overset{\overset{\text{O}}{\|}}{\underset{\underset{\text{H}}{|}}{\text{C}}}-\text{C}- + \text{R}-\text{X} \xrightarrow[\text{base}]{} -\overset{\overset{\text{O}}{\|}}{\text{C}}-\overset{}{\underset{\underset{\text{R}}{|}}{\text{C}}}-$$

R = fragment to be attached at node adjacent to ketone.
X and base = variables to be adjusted depending on the attributes of R and the ketone.

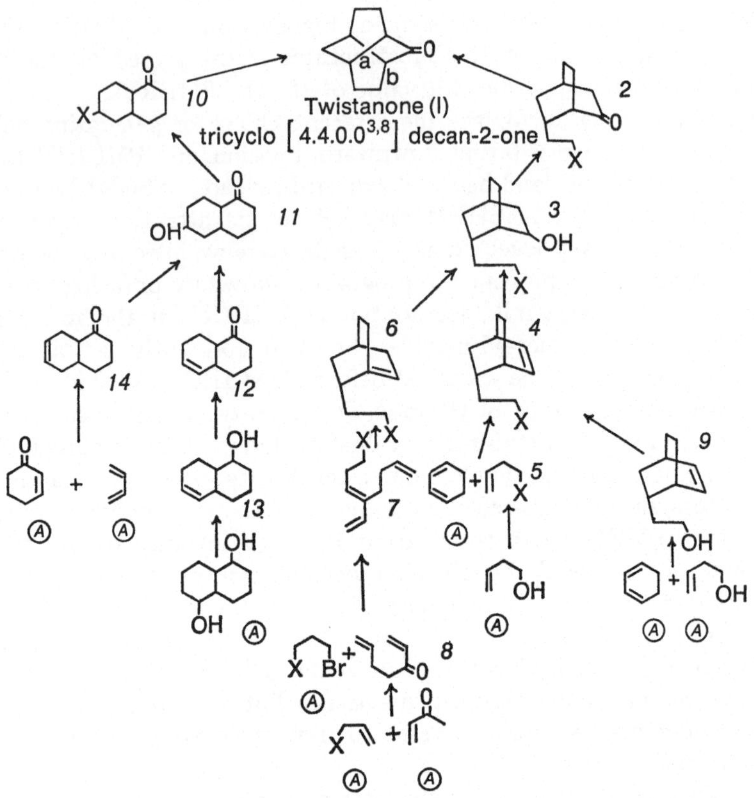

Fig. 6. Synthesis routes discovered by SYNCHEM for twistanone after approximately twenty minutes of search time. Structures labeled "A" were found on the list of available compounds

there are a number of ketone schema in the reaction library, the program selected alkylation because (working backward) it involved breaking a carbon-carbon bond. As pointed out earlier, bond-breaking schema generally result in simpler subgoals with relatively higher values for CPMERIT, and hence, of SGMERIT.

The symmetries of the twistanone molecule are such that two subgoals are generated by the alkylation schema. Cleavage of bond (a) in *1* gives the decalin structure *10*, whereas cleavage of bond (b) yields the bicyclo-[2.2.2] octane structure *2*. The decalin structure is common in organic chemistry, and many derivatives are readily synthesized or are commercially available. The program easily found the two "acceptable" syntheses exhibited.

134

The bicyclo[2.2.2] octane structure is not as common. It is often, and usually most conveniently, prepared by a Diels-Alder reaction[b] using a 1,3cyclohexadiene as the diene component. SYNCHEM, too, followed this route, producing several "acceptable" syntheses from commercially available starting materials. On the other hand, subgoal 6, which appears in the interesting and superficially acceptable route through subgoals 8—7—6—3—2—1, contains a double bond that is very unstable. It is in fact doubtful whether that structure could exist for any length of time at room temperature. This is, however, a relatively easy type of invalid subgoal to prune from the tree when this class of heuristic is programmed into SYNCHEM. Also, SYNCHEM synthesizes Compound 9 by a single Diels-Alder reaction using 1,3-cyclohexadiene and 3-butenyl alcohol. As it stands, this reaction is not really favorable, although it is conceptually correct. Here, too, additional heuristics would enable the program to choose the correct path.

It is most interesting to note that in addition to developing synthesis routes for the bicyclo[2.2.2] octane structure 2 through a Diels-Alder reaction, the program also generated the ketones listed as SG 7 and SG 8 in Fig. 5 (the second page of the search tree containing subgoals developed for structure 2). These molecules, generated by a second alkylation of the bicyclic ketone, have never been employed as intermediates in the synthesis of the twistane structure, but it is believed that SG 8 in particular stands a reasonable chance of providing a successful new route to the synthesis of twistane and its derivatives.

A major deficiency of our system as it is presently constituted is its neglect of stereochemistry, which accounts for the orientation of atoms in space. Compound 10, for example, as displayed in Fig. 6, could represent eight different stereo-isomers. Only two of these eight isomers will lead to the desired goal, twistanone. Until SYNCHEM is able to recognize the stereochemistry of complex molecules, all of its results must be viewed with suspicion.

Considerations of stereochemistry were not included in our first explorations with SYNCHEM because we did not forsee at the outset of this research that our initial efforts would take us so far (from the points of view of complexity of structures manipulated) so fast. We therefore elected to avoid the added complication of spacial variable processing until we had

---

[b] Olefin chapter Reaction Schema 1 for Diels-Alder reaction.

1,3 cyclohexadiene

mastered the more elementary transformations and manipulations. We were surprised to learn that within the framework of our earliest conceptualization of SYNCHEM, we could execute transformations as complex as those demanded by, say, the Diels-Alder schema, and correctly deal with structures as complex as the polycyclic twistane ring system. Our internal representation of the molecular structure has been designed to accommodate readily to stereochemistry, however, and the experienced computer user will recognize that the addition of the further dimension of *bond angle* to our data base and transformation routines is a straightforward, if rather messy problem. At the moment, we lean towards dealing with stereochemistry by post-processing, rather than by modification of existing transformation routines. That is, after a set of synthetic routes has been generated without regard to the orientation of atoms in space, an editing routine will fill in the missing bond angles, and reject those pathways that violate the stereochemical constraints of the reaction schema that are used.

We promised earlier that we would indicate where intervention was necessary before the complete set of syntheses in Fig. 6 were forthcoming. The identification numbers on the figure indicate the order in which these compounds were developed. (Other compounds which were selected as generating subgoals, but whose subtrees did not figure in any of the completed routes are not represented in the serial ordering.) After finding the acceptable synthetic routes *5—4—3—2—1* and *9—4—3—2—1*, and the unsatisfactory route *8—7—6—3—2—1*, SYNCHEM selected another subgoal on that branch of the tree and began to follow it down a very rapidly-proliferating but sterile subtree. Since the progression of values for SGMERIT gave no hint of SYNCHEM's imminent disenchantment with that branch, the search was manually terminated, and the program was started on the decalin structure *10*, to produce the synthesis routes *13—12—11—10—1*, and *14—11—10—1*. While we have not yet been able to find a non-destructive modification of the search algorithm which will enable the program to find all of the routes in Fig. 6 without intervention and without excessive barren exploration, it is clear (if not especially comforting) that given enough computation time, the depth-limiting factor in the search algorithm as it is presently programmed would ultimately force SYNCHEM to generate the complete tree.

III. Some Further Excerpts from the Search Trace

As we remarked above, it is certainly not always the case that every reaction schema in a chapter of the reaction library passes the A-test for that chapter and is accepted for further processing. Fig. 7 illustrates the elimination of schema by A-test and, it so happened here, by N-test as well. The arrows

```
CURRENT CHAPTER IS ALDEHYDE                  ↓    ↓↓

ABITS OF GOAL        0110000000000000000000110000001100000000000
ATEST FOR SCHEMA  1  0000000000000000000000000000000001000000000
ATEST FOR SCHEMA  2  0000010000000000000001001000001100000101101  - DISCARDED
ATEST FOR SCHEMA  4  0000000000000000000000000000111100000000100  - DISCARDED

ATEST SUCCEEDED

ELIGIBLE_SCHEMA          '100000000000000000'B



CURRENT SCHEMA NUMBER IS                                   1

GOAL PATTERN  $1VH

SUBGOAL PATTERN  $1:1C

ALDEHYDE        ** OXIDATION OF ALCOHOL ** USE DMSO/DCC OR DICHROMATE IN H2SO4
EASE 10  NSTEPS   1


ALCO1                 NOT SUCCESSFUL/ 2


ALCO2                 NOT SUCCESSFUL/ 2
*** TEST ANSWERED YES
```

Fig. 7. Schema rejection by A-test and N-test

indicate where conflicts occur; bits 21, 28, and 29 represent the olefin bond, aldehyde, and alcohol, respectively. Schema 2 was discarded because the generating subgoal compound has at least one olefin bond and at least one aldehyde group (either attribute would have been sufficient to cause rejection). Schema 4 is discarded when the generating subgoal contains an aldehyde group, and also when it contains an alcohol. Schema 1, which fails outright only in the presence of an ether group (bit 30) is eligible for further development. But here, the discovery by N-test that a secondary alcohol group is present in the generating subgoal (ALCO2 ***TEST ANSWERED YES) causes this reaction (OXIDATION OF ALCOHOL) to be rejected as well. As in the case of ketone formation by oxidation of an alcohol discussed earlier, a tertiary alcohol cannot be oxidized to a carbonyl group, and so the presence of bit 29 in the ABITS string is not alone sufficient reason to discard the schema.

A somewhat more complex kind of N-test is illustrated in Fig. 8, an excerpt from the trace through the halide library. Schema 1, for the halogenation of a hydroxy group, can proceed through several different reagents of varying acidity (although not every reagent will synthesize every halide). Execution of the test for acid sensitivity here adds to the attribute table extension a list of all functionalities in the generating subgoal that are attacked by acids, together with an ad hoc estimate of the maximum reagent

137

```
CURRENT SCHEMA NUMBER IS                                    1

GCAL PATTERN  $2X$4$6J                                             CURRENT CHAPTER IS HALIDE

SLBGOAL PATTERN  $2X$4$6Q

FROM ALCOHOLS
SEXIT   2
FEXIT   2
EASE    C.89941

ACID SENSITIVE                                  3 -1C3    0
FOR 83  RESULT = CCCCCCCCOOOCCCCCCCCCCCCCCCCCOCOOO
TEST NOT DEFINED
TEST NOT DEFINED
BRANCHING TO ALCOHOL
FOR 43  RESULT = C1COCCCOCOGCCCCCCCCCCCCCCCCOCOCO
            42              SENSITIVE#            5.5C00
METHOD   1   TCO SENSITIVE:  CANCELLED
METHOD   3   TCO SENSITIVE:  CANCELLED
EASE=   C.89941
THERE EXIST   2  METHODS
METHOD      EASE              NAME
    4   0.89941  CONVERT ALCOHOL TO TOSYLATE AND DISPLACE WITH NAX IN POLAR SOLVENT
    2   0.69922  USING TRICHLOROMETHYL_TRIPHENYL PHOSPHONIUM CHLORIDE
*** CHANGE EASE VALUE
NEASE=   C.89941.

CHOOSING CELL 32 OF GOAL FOR MATCHING,  3 MATCHES FOUND.
L66 A BTJ AC E1:1Q

SUBGOAL GENERATED L66 A BTJ AQ E1:1C

NEXT SET CF SUBGOALS FOR THE SAME SCHEMA
L66 A BTJ AQ E1:1Q

SUBGOAL GENERATED L66 A BTJ AQ E1:1C
** WHOLE SET ELIMINATED

NEXT SET CF SUBGOALS FOR THE SAME SCHEMA
L66 A BTJ AC E1:1Q
SUBGOAL GENERATED L66 A BTJ AQ E1:1C
** WHOLE SET ELIMINATED

L66 A BTJ AQ E1:1C  NOT AVAILABLE
```

Fig. 8. Acid sensitivity scale N-test

acidity (on an arbitrary scale of 10) that each such group will tolerate. The test returns the maximum acidity (5.5 in the example) that will be tolerated by *all* functionalities in the generating subgoal. In the excerpt from Schema 1, methods 1 and 3 require reagents of acidity greater than 5.5, and are discarded. Methods 2 and 4 remain, and are listed as possibilities. Of these, method 4 has the highest ad hoc EASE (for technical reasons, the scale of EASE is normalized to a maximum of 1.0 in the halide library) and is the method of choice, since it is suitable for any halide. Method 2, which proceeds through a rather less acid reagent would have been the sole surviving method had the acid sensitivity of the generating subgoal been somewhat greater. It is suitable only for the synthesis of chlorides, however, and would have been discarded if the syntheme being processed had been a halide other than a chloride or the arbitrary halide designation. If, on the other hand, the syntheme were a chloride or arbitrary halide. EASE would have been adjusted downwards as specified, and, in the latter case, the arbitrary halide would have been set to chloride in that branch of the search tree.

PARENT = CMPN   9

CMPN 15 = L66 A E ALTJ C1:1J
CPMERIT = 5.3125

PST LEVEL =   4, NUMBER CF SUBGOALS =   9, NUMBER OF CERIVEC CCMPOUNDS =   11.

TSC CELL =   6, REACTICN SCFEMA = 1, CHAPTER = CLEFIN   BCNC

NUMBER OF SUBGCALS GENERATED USING RS 1 =   2.

SG 1, EASE =   9, NSTEPS = 1, CCNFIC =   10, SGMERIT = 4.6875

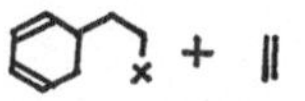

CMPN 17 = 1U1
CPMERIT = 8.75CC

CMPN 18 = L6U CUTJ E1:1J
CPMERIT = 6.COOC

SG 2, EASE =   9, NSTEPS = 1, CCNFIC = ,10, SGMERIT = 6.9375

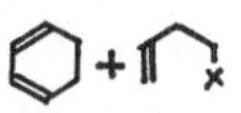

CMPN 19 = J1:1:1L1
CPMERIT = 7.75OC

CMPN 20 = L6U CUTJ
CPMERIT = 1C.COCC (CN LIST CF AVAILABLE CCMPOUNDS)

TSC CELL =   6, REACTICN SCHEMA = 2, CHAPTER = OLEFIN   BCNC

NUMBER OF SUBGCALS GENERATED USING RS 2 =   1.

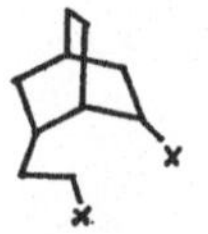

SG 3, EASE =   8, NSTEPS = 1, CCNFIC =   10, SGMERIT = 4.2500

CMPN 21 = L66 A BTJ AJ E1:1J
CPMERIT = 5.3125

TSC CELL =   6, REACTICN SCHEMA = 3, CHAPTER = OLEFIN   BONC

NUMBER OF SUBGCALS GENERATED USING RS 3 =   1.

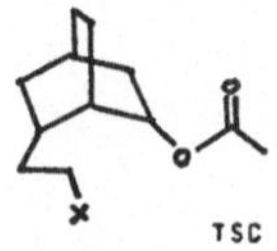

SG 4, EASE =   9, NSTEPS = 1, CCNFIC =   10, SGMERIT = 4.6875

CMPN 22 = L66 A BTJ ACV1 E1:1J
CPMERIT = 5.25CC

TSC CELL =   6, REACTICN SCFEMA = 5, CHAPTER = CLEFIN   BCNC

NUMBER OF SUBGCALS GENERATED USING RS 5 =   1.

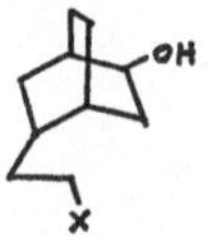

SG 5, EASE =   8, NSTEPS = 1, CCNFIC =   10, SGMERIT = 4.2500

CMPN 24 = L66 A BTJ AQ D1:1J
CPMERIT = 5.3125

TSD CELL =   8, REACTICN SCHEMA = 2, CHAPTER = CLEFIN   BCNC

NUMBER OF SUBGCALS GENERATED USING RS 2 =   1.

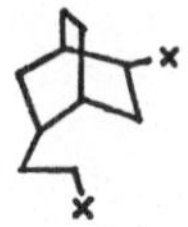

SG 6, EASE =   8, NSTEPS = 1, CCNFIC =   10, SGMERIT = 4.2500

CMPN 25 = L66 A BTJ AJ D1:1J
CPMERIT = 5.3125

TSD CELL =   8, REACTICN SCFEMA = 3, CHAPTER = CLEFIN   BONC

NUMBER OF SUBGCALS GENERATED USING RS 3 =   1.

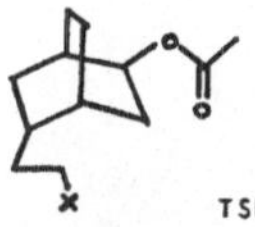

SG 7, EASE =   9, NSTEPS = 1, CCNFIC =   10, SGMERIT = 4.6875

CMPN 26 = L66 A BTJ ACV1 C1:1J
CPMERIT = 5.25CC

TSC CELL = 28, REACTICN SCHEMA = 1, CHAPTER = HALIDE

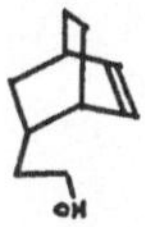

NUMBER OF SUBGCALS GENERATED USING RS 1 =   1.

SG 8, EASE =   9, NSTEPS = 1, CCNFIC =   10, SGMERIT = 4.7500

CMPN 28 = L66 A B AUTJ D1:1Q
CPMERIT = 5.3125

TSC CELL = 28, REACTICN SCFEMA = 3, CHAPTER = HALIDE

NUMBER OF SUBGCALS GENERATED USING RS 3 =   1.

SG 9, EASE =   8, NSTEPS = 1, CCNFIC =   10, SGMERIT = 4.2500

CMPN 29 = L66 A B AUTJ D1U1
CPMERIT = 5.3125

Fig. 9. Partial terminus of first twistanone synthesis route

139

CMPN 19 = J1:1:1U1
CPMERIT = 7.75C0

PST LEVEL = 5, NUMBER OF SUBGOALS = 10, NUMBER OF DERIVED CCMPOUNDS = 12.

TSD CELL = 12, REACTICN SCHEMA = 2, CHAPTER = CLEFIN   BOND

NUMBER CF SLBGCALS GENERATED USING RS 2 = 1.

SG 1, EASE =   8, NSTEPS = 1, CCNFID = 10, SGMERIT = 8.0000

CMPN 30 = J1:2:1J
CPMERIT = 1C.COCC (CN LIST OF AVAILAELE CCMPOUNDS)

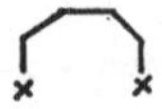

TSD CELL = 12, REACTICN SCHEMA = 3, CHAPTER = CLEFIN   BOND

NUMBER OF SLBGCALS GENERATED USING RS 3 = 1.

SG 2, EASE =   9, NSTEPS = 1, CCNFID = 10, SGMERIT = 6.8125

CMPN 31 = J1:2:1CV1
CPMERIT = 7.6250

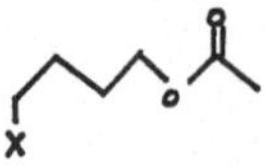

TSC CELL = 12, REACTICN SCHEMA = 4, CHAPTER = CLEFIN   BOND

NUMBER OF SLBGCALS GENERATED USING RS 4 = 1.

SG 3, EASE =   8, NSTEPS = 3, CCNFID = 10, SGMERIT = 6.2500

CMPN 32 = E1
CPMERIT = 1C.COCC (CN LIST OF AVAILAELE CCMPOUNDS)

CMPN 33 = VH1:1J
CPMERIT = 7.8125

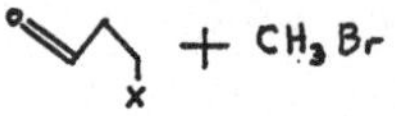

TSD CELL = 12, REACTICN SCHEMA = 5, CHAPTER = CLEFIN   BCND

NUMBER GF SLBGCALS GENERATED USING RS 5 = 1.

SG 4, EASE =   8, NSTEPS = 1, CCNFID = 10, SGMERIT = 6.1875

CMPN 34 = QY1G1:1J
CPMERIT = 7.750C

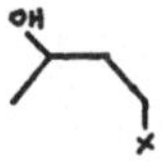

TSD CELL = 14, REACTICN SCHEMA = 2, CHAPTER = CLEFIN   BCND

NUMBER OF SLBGCALS GENERATED USING RS 2 = 1.

SG 5, EASE =   8, NSTEPS = 1, CCNFID = 10, SGMFRIT = 8.0000

CMPN 35 = JY1G1:1J
CPMERIT = 1C.COOC (CN LIST OF AVAILAELE CCMPOUNDS)

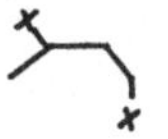

TSD CELL = 14, REACTICN SCHEMA = 3, CHAPTER = OLEFIN   BOND

NUMBER CF SLBGCALS GENERATED USING RS 3 = 1.

SG 6, EASE =   9, NSTEPS = 1, CCNFID = 10, SGMERIT = 6.8125

CMPN 36 = J1:1Y1G0V1
CPMERIT = 7.6250

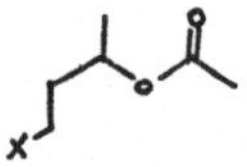

TSD CELL = 14, REACTICN SCHEMA = 4, CHAPTER = CLEFIN   BCND

NUMBER OF SLBGCALS GENERATED USING RS 4 = 1.

SG 7, EASE =   8, NSTEPS = 3, CCNFID = 10, SGMERIT = 8.0000

CMPN 37 = J1:1:1E
CPMERIT = 1C.COOC (CN LIST OF AVAILAELE CCMPOUNDS)

CMPN 38 = VHH
CPMERIT = 1C.COOC (CN LIST CF AVAILAELE CCMPOUNDS)

TSC CELL = 14, REACTICN SCHEMA = 5, CHAPTER = CLEFIN   BOND

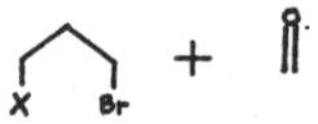

NUMBER CF SLBGCALS GENERATED USING RS 5 = 1.

SG 8, EASE =   8, NSTEPS = 1, CCNFID = 10, SGMERIT = 6.1875

CMPN 39 = C1:2:1J
CPMERIT = 7.750C

TSC CELL = 6, REACTICN SCHEMA = 1, CHAPTER = HALIDE

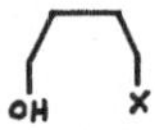

NUMBER OF SLBGCALS GENERATED USING RS 1 = 1.

SG 9, EASE =   9, NSTEPS = 1, CCNFID = 10, SGMERIT = 9.0000

CMPN 40 = Q1:1:1U1
CPMERIT = 1C.COOC (CN LIST OF AVAILAELE CCMPOUNDS)

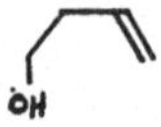

TSD CELL = 6, REACTICN SCHEMA = 3, CHAPTER = HALIDE

NUMBER OF SUBGCALS GENERATED USING RS 3 = 1.

SG1C, EASE =   8, NSTEPS = 1, CCNFIC = 10, SGMERIT = 8.0000

CMPN 41 = 1U1:1U1
CPMERIT = 1C.COOC (CN LIST OF AVAILAELE CCMPOUNDS)

Fig. 10. Completion of terminus for first twistanone synthesis

140

SYMBOL TABLE

CP #	SYMB	INSTANCES	DEVELOPMENT SITE SG #	CF	CP #	CATALOG#	WIS NAME
1	GG1G	1	(GOAL CCMFCLNC)				L666/BI/EJ A A 2BF J AVTJ
2	006C	1					L666/CH B B 2AC JTJ BQ
3	C0A8	1					L66 ATJ CCN G-MG-J
4	CCFG	1					L66 BVTJ HJ
5	0130	1	4		1		L66 A B AVTJ E1:1J
6	0178	1					L66 A B AVTJ E1:1Q
7	01CG	1					L66 A B AVTJ E1:1VQ
8	C208	1					L66 A B AVTJ E1L1
9	C250	1	4		5		L66 A BTJ AC E1:1J
10	C2SE	1					L6TJ ACN B1:1J D1-MG-J
11	C2E0	1					L6TJ A-MG-J B1:1J D1CN
12	0328	1					L6VTJ C1YJ1:1J
13	036E	1					L6VTJ C1:1J E1:1J
14	C3BC	1					L66 A B AU- FTJ E1:1J
15	C3F9	1	2		9		L66 A B ALTJ D1:1J
16	C44C	1					L66 A BTJ AQ E1:1Q
17	0488	1					1U1
18	04C8	1					L6U CUTJ E1:1J
19	C5C8	1	2		15		J1:1:1U1
20	C548	1				C1CCCG5	L6U CUTJ
21	C5E8	1					L66 A BTJ AJ E1:1J
22	05CC	1					L66 A BTJ ACV1 E1:1J
23	C618	0					L6TJ A1E B1:1J DVH
24	C66C	1					L66 A BTJ AC D1:1J
25	06A8	1					L66 A BTJ AJ D1:1J
26	C6FC	1					L66 A BTJ ACV1 D1:1J
27	0738	0					L6TJ AVH B1:1J D1E
28	C78C	1					L66 A B AUTJ D1:1Q
29	07C8	1					L66 A B AUTJ C1L1
30	C810	1				14CEC5	J1:2:1J
31	C85C	1					J1:2:1OV1
32	C85C	1				YCCOC02	E1
33	C8C0	1					VH1:1J
34	C91C	1					CY1E1:1J
35	C95C	1				156574	JY1E1:1J
36	C99C	1					J1:1Y1&OV1
37	C9CC	1				125903	J1:1:1E
38	CA1C	1				FC15587	VHH
39	CA5C	1					C1:2:1J
40	CA90	1				11C361	C1:1:1U1
41	CACC	1				YCCC031	1U1:1U1

Fig. 11. Symbol table after the first synthesis was completed

We conclude our discussion of the twistanone search trace with the two pages of the tree that contain the terminus of the first synthesis route in the tree. It will be seen that subgoal 2, derived through a Diels-Alder schema, splits the generating subgoal into a cyclic compound which is found on the shelf library (1,3-cyclohexadiene) and a simple linear halogenated alkene, which is assigned a relatively high value for CPMERIT (Fig. 9). The resulting high value for SGMERIT makes subgoal 2 the choice for the next generating cycle (Fig. 10), where five routes terminate in the shelf library. Two of these (subgoals 1 and 5) synthesize the olefin bond by dehydrohalogenation, one (subgoal 7) synthesizes the olefin bond with a Wittig reaction (notice that NSTEPS = 3 for the Wittig), one (subgoal 9) synthesizes the halide from the alcohol, and one (subgoal 10) synthesizes the halide from the olefin bond. SYNCHEM'S preferred route (*i.e.*, with the highest value for SGMERIT) is subgoal 9, synthesizing the 3-butenyl halide from 3-butenyl alcohol. Fig. 11 is a printout of the symbol table at this stage of the search.

141

IV. Other Results

Earlier, we gave "several dozen" as the number of synthesis problems SYNCHEM had solved to date. It is impossible to be more precise than this, for organic structures which taxed the limits of SYNCHEM's competence at earlier stages in its development often reappear in identical or closely related form in our more recent work as subgoals for a far more complex target compound. Should the earlier synthesis still be counted among our successes, such as they are? And if so, should not every synthetic route to every intermediate compound in our later syntheses also be counted separately?

Rather than wrestle with this issue, which is obviously a moral rather than a scientific one, we have elected to count only the several dozen or so distinctly different kinds of organic structures we have dealt with as separate problems. Since our interpretation of "distinctly different" is highly subjective (we tend to define differences in terms of SYNCHEM's behavior), our answer to the question "How many?" must always remain vague.

One of these results, however, deserves special mention. The target molecule was Vitamin A, an organic structure of obvious bio-chemical interest, and one for which there exist a number of well-known synthetic routes. In this case, the program, guided by ostensibly reasonable albeit incomplete heuristics, produced a collection of relatively poor syntheses (Fig. 12). SYNCHEM started out correctly enough by recognizing that a Wittig reaction for preparation of the olefin bond provided a likely approach to the problem. This being the case, the program, again correctly, preferred to prepare the ester (2) by the Wittig reaction, rather than the target molecule directly, observing that the hydroxy group in Vitamin A renders the direct Wittig (the left branch of the tree sprouting from the target molecule) less favorable. SYNCHEM performed well, too, in recognizing that the Wittig schema applied to the olefin bond indicated by the arrow in the target molecule (and also in the ester, 2) generates as one of the two subgoal reactants the readily available compound β-Ionone, which can provide, prefabricated, much of the complex functionality required for the target molecule. Here, however, SYNCHEM lacked the heuristic sophistication to forsee that immediate application of the Wittig schema to the indicated olefin syntheme to "separate out" the β-Ionone would produce a subgoal (3, in Fig. 12) of such synthetic complexity as to confound its resources at a later stage of the search. The relatively reactive and varied functionalities in the molecules generated as subgoals for the synthesis of Compound 3 make them undesireable reactants for a procedure with so many steps. A far better synthetic route for Vitamin A (following the well-known procedures in the literature) calls for building the target molecule little by little, starting with β-Ionone, and adding a few carbons at a time to the chain. (From the back-

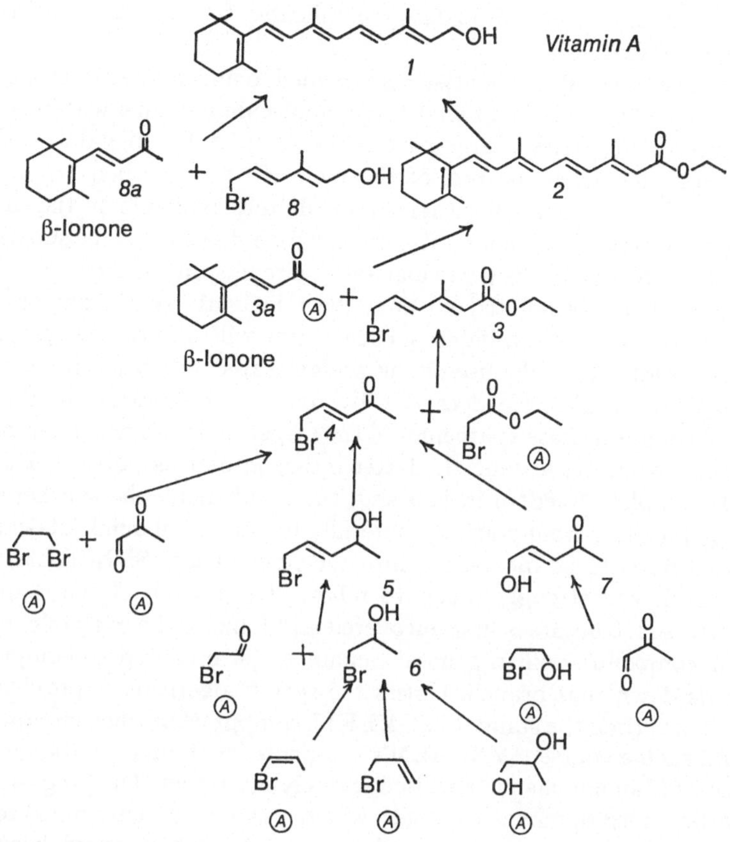

Fig. 12. Synthesis routes discovered by SYNCHEM for Vitamin A after approximately ten minutes of search time

ward point of view of the search tree, this means that the right-most olefin bond in 2 should have been selected for development by the Wittig schema, then the properly prepared rightmost olefin bond in the ring-containing fragment produced by the first Wittig, and finally the rightmost olefin bond in the ring-containing fragment generated by the second Wittig to "separate out" the β-Ionone.)

A postmortem on the Vitamin A output resulted in the formulation of an interesting new heuristic which, we feel, will enable SYNCHEM to duplicate an important kind of insightful behavior exhibited by chemists faced with the problem of finding synthetic routes for very complicated organic molecules. Although it was invented to save us from being embarrassed by Vitamin A, the heuristic is clearly of general applicability for organic syn-

143

thesis discovery, and is probably generalizable to other areas of heuristic problem solving.

Briefly expressed, the heuristic is intended to inform the search algorithm when a complex multifunctional target molecule contains within it a substructure which carries a large part of the complex functionality such that, when separated from the rest of the molecule by a carbon-carbon bond-splitting schema, the substructure is immediately available on the shelf list. This is seen to be the case for Vitamin A, where β-Ionone is the substructure in question. Noticing this particular set of circumstances, the organic chemist (who, it should be pointed out, must be sufficiently skillful and experienced to recognize the available substructure) will almost always plan his synthetic routes to make use of the prefabricated substructure in building the target molecules. He does this by focusing his transformations and manipulations on those synthemes of the target molecule which are not part of the substructure, leaving the latter intact insofar as possible. He allows himself complete freedom in choosing the point along the synthetic route (looking forward, from starting materials to goal compound) at which the substructure is to be introduced into the preparation. SYNCHEM, on the other hand, will strongly prefer to follow that branch of the search tree where the substructure is first uncovered and found to be available, for a far simpler compound will in general remain to be prepared to complete the synthesis down that branch. Unless the search algorithm is provided with means to override the standard SGMERIT computation when circumstances demand such a course, SYNCHEM's prospects for simulating the intelligent behavior of human chemists will be severely restricted. The program would always be forced to pursue that path which was naively determined to be the obviously simplest one, missing all those routes which avoid hazards by taking a detour.

That SYNCHEM already behaves this way to a limited extent is evident in the first level of the Vitamin A synthesis tree, where the algorithm elects to develop the apparently more complex ester *2* rather than the obviously simpler compound *8* which remains after the β-Ionone, *8a*, has been separated out. An N-test for the Wittig reaction (olefin chapter, Schema 4) which checks for the presence of a hydroxy group in the target molecule enabled SYNCHEM to make the correct choice of subgoal. A positive result for the test causes the search algorithm to boost up the value of SGMERIT for the alternate branch of the tree that synthesizes the hydroxy group from the ester, *providing that the ester subgoal passed its own A-, N-, and F-tests to earn a place on the tree*. Thus, if a reasonably good transformation from the ester back to the alcohol is not available, SYNCHEM will follow its normal preference and choose the subgoal-simplifying reaction.

Programming the more general heuristic procedure for enabling SYN-CHEM to override its naive choice of preferred subgoal (we call it the ASM, for Available Substructure Maintenance, heuristic, since its purpose is to keep the available substructure intact while allowing free development of the other synthemes in the target molecule) has been a much more difficult task. While it is easy enough to define the function of the heuristic conceptually, its realization as a collection of subroutines within the problem-solving environment of organic chemistry has proved to be a challenging exercise in program design, and at the time of writing this report, was not yet ready for incorporation into SYNCHEM. With the ASM heuristic a working part of the search algorithm, the development of the Vitamin A syntheses will be modified in the following way. Having recognized β-Ionone to be a complex but available substructure of the target molecule, the program identifies those atoms and synthemes of Vitamin A which are contributed by the β-Ionone. The purposes of the heuristic are achieved it two ways. First, the value of SGMERIT is penalized for all subgoals developed for the identified synthemes, and second, the value of CPMERIT for all other generated subgoal compounds is computed as if the identified atoms and synthemes were not part of the structure. The latter procedure informs the search algorithm that subgoal compounds which contain the intact substructure of β-Ionone are less complicated than they might appear to be, and in fact, no more complex than the part that remains after the β-Ionone is separated out. While the Vitamin A synthesis has been used to illustrate the function of the ASM heuristic, it is clear that its application is completely general in the context of organic synthesis discovery.

V. Conclusion

We are by no means so sanguine that we claim more for our achievements to date than that they make initial inroads into a problem domain that offers rich rewards to its ultimate conquerers. We do feel, however, that they are of sufficient substance to warrant the conclusion that our approach to the problem of computer-designed organic synthesis can yield a system of real value for the synthetic chemist, perhaps useful enough to be routinely called upon in organic chemical investigations. The reasonable computation times achieved on our IBM System 360/67, averaging about 30 sec for each cycle of subgoal generation, encourage us to believe that our ultimate goals are well within the capability of contemporary computer technology.

Even in its present rudimentary stage of development, SYNCHEM is often successful in "roughing out" synthetic routes that are conceptually correct. It is clear that an enormous amount of work remains to be done on our reaction library, both extending and refining it, before the details are set to right. From the start, SYNCHEM enjoys certain advantages over its human counterpart, for it is not hindered in planning a synthesis by the difficulty of visualizing three-dimensional structures. For example, it may not be immediately obvious to the chemist that structures *a* and *b* in Fig. 13 are identical, or, in viewing structure *a*, that it possesses a plane of symmetry. To the program, the properties of a given structure are computed from the TSD. They are, of course, independent of the organization of the data, provided that the connectivity is expressed correctly.

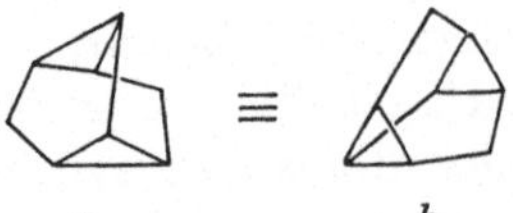

$$a \equiv b$$

a   *b*

Fig. 13

By the same token, SYNCHEM can rapidly identify structural features in organic molecules that may not be immediately obvious to chemists. Tricyclo[2.2.2.0^{2,6}]octene is a complex organic molecule whose synthesis could pose problems for the inexperienced chemist. The program immediately recognizes the structural features necessary for a Diels-Alder reaction, and suggests that this molecule might be synthesized in one step from the considerably simpler 6-vinyl-1,3-cyclohexadiene (Fig. 14). In fact, because the computer is immune to the bias of experience that the chemist might find difficult to avoid when the molecule he must synthesize seems similar to one whose synthesis he knows, it would not be surprising if SYNCHEM were to find novel, and possibly superior, routes for the synthesis of well-known compounds. It is for this reason that completely self-guided synthesis discovery programs, as opposed to user-guided interactive ones, are of genuine interest to chemists as well as to computer scientists engaged in artificial intelligence research.

6-Vinyl-1,3-cyclohexadiene   Tricyclo[2.2.2.0$^{2.6}$]octene

Fig. 14

We were able to master the mechanics of synthesis discovery relatively quickly because we were able to profit from much of the published work in heuristic problem solving that makes up a substantial part of the early literature of artificial intelligence. The computer's present lack of brilliance in organic chemistry itself is, we feel, correctable. Even in its currently naive stage of development, the program and data structures provide for the consideration of many factors of importance in synthetic chemistry, but for which the heuristic decision functions have yet to be programmed. For example, we have not yet applied the heuristic coefficients stored in the machine's internal representation of the TSD which are intended to suggest the degree to which electron-withdrawing or electron-donating tendencies of specific functionalities are transmitted to neighboring carbon atoms at varying distances from the activating site. Indeed, once the ASM heuristic is a working part of the SYNCHEM search algorithm, some recent work of Feldmann, Heller, Shapiro, and Heller [12], offers intriguing possibilities. As part of a collection of programs prepared for the National Institutes of Health, and intended to provide extensive chemical documentation and information retrieval services in an interactive computer system, they have written a series of subroutines called SSS (for Substructure Search) and RSS (for Rapid Structure Search) which take as input the specification of an organic substructure, and deliver as output those items on a very large file of organic compounds that have the input substructure imbedded within it. It is easy to see how one might combine substructure search with the ASM heuristic to guide the development of the synthesis search tree to terminate on starting materials that provide the maximum prefabricated functionality for the target molecule, and hence, presumably, the most efficient syntheses.

Although SYNCHEM cannot generate intermediate organic structures that violate the rules of bond multiplicity, it can and does (as we have seen in the twistanone trace) generate molecules that cannot exist for some other reason, excessive bond strain, for example, or energetic instability. Here, the earlier work cited in theorem proving offers few clues to how one should proceed. For while the geometry program, for example, could test the validity of a generated expression by interpreting it in a model of the formal system (in that case, a diagram), no such simple device is available to us now to determine whether a molecular structure can in fact exist under the conditions specified for the synthesis. One can, of course, make use of chemical tables if the molecular structure in question is a known one and the information required is in the literature. To restrict the program to such intermediate compounds, however, and have it assume that untabulated structures cannot be valid subgoals, would severely limit the range of possible syntheses that might be discovered. Failing a response from a table lookup, then the validity of a computer-created molecule as a subgoal might be determined in one of two ways. First, we may press into service whatever results of physical chemistry we can systematize, generalize, and formalize sufficiently to incorporate into tree-pruning heuristics. These could be as simple-minded as a program which estimates melting and boiling points from the molecular weight and comparisons of the carbon skeleton of the unknown compound with a standard scale of materials, or as sophisticated as an application of some form of Wahl's BISON program [13], which, by solving in approximation a quantum-mechanical model of the electronic structure of the molecule in question, is able to compute electron bonding energies, charge distributions, and estimates of certain macroscopic physical properties of the compound. Falling towards the lower end of the scale in sophistication are heuristics of the kind that could eliminate compound 6 from the twistanone search tree (Fig. 6), which is based upon the empirically observed instability of a double bond to a polycyclic ring junction.

A second possible approach to the question of determining subgoal validity is that followed so often by the chemist; do (or in the case of our program, suggest) an experiment to supply the missing information.

Thus far, we have confined our attention to solution generation, the analytic search phase of a complete synthesis discovery system. Once the analytic search routines have accumulated a satisfactory set of proposed syntheses for the target molecule, solution evaluation procedures must now be applied to sort out the good and usable routes from the merely reasonable ones. Of two proposed syntheses, each requiring the same number of steps from essentially similar starting materials, it is quite possible that one might produce a high yield of the target molecule in the pure state, while the other might result in negligible yield in a difficult-to-separate mixture of by-products. Our next step, then, must be to invert the analytic search (backwards, from goal to starting point) used to discover the proposed solutions, and adopt a synthetic approach (forwards, from starting materials to goal), literally doing chemistry by computer. In this synthetic reconstruction phase, each designated reaction of a proposed synthesis is simulated in the computer to determine what happens in addition to the interaction for which that mechanism was selected. Thus, if a given syntheme is to be prepared by reacting specified functional groups on two starting compounds, we must predict which other functionalities, if any, in the starting compounds will interact at the same time, and what byproducts will thereby enter the arena, and in what concentrations. Or, if the specified groups can interact in more than one way, we must estimate the relative yields of the competing reactions. Some of these judgements are of course part of the chemical heuristic tests which are applied during the process of subgoal generation, but here we wish to make quantitative, rather than qualitative determinations. Once again, the program must make use of tables, physical-chemical models, heuristics, and occasional requests for experimental data to aid in estimating relative yields and byproducts, and to estimate separation efficiencies when mixtures are produced. The overall evaluation of the relative merit of each of the proposed syntheses will be a composite of the propagated yield and ease of execution of each step of a procedure (including separation and purification stages), modified by the starting constraints on the problem. It goes without saying that the final judgement remains the prerogative of SYNCHEM's chemist users.

We do not underestimate the difficulty of rendering the synthetic reconstruction phase of our system in software. Nevertheless, those aspects of the problem that fall within the domain of computer science seem well enough defined, and solvable with no more than the usual portion of pain and suffering. It is less clear to us at the moment just how we will introduce a measure of the chemist's wisdom into SYNCHEM's evaluations. But it is in the nature of artificial intelligence research that such behavior seems to evolve from the accumulation of many obvious little solutions to many immediate little problems. We have no reason to believe that anything other than this will be the case for SYNCHEM.

Acknowledgments. We thank Mrs. Marcia Chachamovitz Fortes for her important contributions to the TSD-WLN conversion programs. This work was supported largely by the Research Foundation of the State University of New York, with assistance from a NSF Computer Science Program Development Grant.

VI. References

1) Amarel, S.: Problem solving and decision making by computer: an overview. Cognition and artificial intelligence: a multiple view (ed. Garvin). Aldine Press 1969.
2) Lederberg, J., Sutherland, G. L., Buchanan, B. G., Feigenbaum, E. A.: A heuristic program for solving a scientific inference problem; summary of motivation and implementation. Theoretical approaches to non-numerical problem solving (eds. Banerji and Mesarovic). Berlin-Heidelberg-New York: Springer 1969.
3) Corey, E. J., Wipke, W. T.: Computer-assisted design of complex organic syntheses. Science *166*, 178 (1969). Wipke, W. T.: A new approach to computer-assisted design of organic syntheses. Proc. Conf. on Cumputers in Chem. Educ. and Res., Northern Ill. Univ., 1971.
4) Ugi, I., Gillespie, P.: Matter preserving synthetic pathways and semi-empirical computer assisted planning of syntheses. Angew. Chem., Internat. Ed. Engl. *10*, 915 (1971).
5) Hendrickson, J. B.: J. Am. Chem. Soc. *93*, 6847 (1971).
6) Corey, E. J., Wipke, W. T., Cramer, R. D., Howe, W. J.: J. Am. Chem. Soc. *94*, 431, 440 (1972).
7) Gelernter, H.: Machine-generated problem-solving graphs. Mathematical Theory of Automata. MRI Symp. Series, Vol. XII. New York: Polytech. Inst. of Brooklyn Press, 1963. — Gelernter, H.: Realization of a geometry theorem proving machine. Proc. of the First. Int. Conf. on Information Proc. UNESCO, published as *Information processing*, Oldenbourg (Munich 1960).
8) Sridharan, N. S., Gelernter, H., Hart, A. J.: Subgoal generation and synthesis search procedures in SYNCHEM, a computer program for the discovery of synthesis routes for organic molecules. In preparation.
9) Smith, E. G.: The Wiswesser line-formula chemical notation. New York: Mc-Graw-Hill 1968.

H. Gelernter *et al.*

10) Whitlock, H. W., Siefken, M. W.: J. Am. Chem. Soc. *90*, 4929 (1968). — Whitlock, H. W.: J. Am. Chem. Soc. *84*, 3412 (1962).
11) Belanger, A., Lambert, Y., Deslongchamps, P.: J. Can. Chem. *47*, 794 (1969). — Gauthier, J., Deslongchamps, P.: J. Can. Chem. *45*, 297 (1967).
12) Feldmann, R. J., Heller, S. R., Shapiro, K. P., Heller, R. S.: An application of interactive computing - a chemical information system. J. Chem. Doc. *12*, 41 (1972). — Meyer, E.: Versatile computer techniques for searching by structural formulas, partial structures, and classes of compounds. Angew. Chem., Intern. Ed. Engl. *9*, 583 (1970).
13) Wahl, A. C., Bertoncini, P., Kaiser, K., Land, R.: BISON: A new instrument for the experimentalist. Intern. J. Quantum Chem. *3* (1970).

Received January 8, 1973

Orientation and Stereoselection

By **K. Fukui**
With 47 figures. 85 pages. 1970
(Topics in Current Chemistry,
Vol. 15/Part 1)
DM 34,–; US $ 12.60

The chemical interaction between
molecular systems is divided into the
Coulomb interaction, the exchange
interaction, the charge-transfer interaction, and the polarization interaction. By way of the charge-transfer
interaction and the polarization interaction, the different electron, configurations come to be mixed with the
initial one. This is a chemical excitation
process. The molecular shape will tend
to change so as to take on a more
stable nuclear configuration. That is to
say, the change in the electronic state
impels the nuclei to rearrange themselves. As the reaction proceeds, each
reactant molecule changes the nuclear
configuration in the direction of
stabilization. Such nuclear rearrangement, on the one hand, is accompanied
by an unstabilization which, like pro-
motion in molecule formation from
atoms, is the principal origin of "activation energy". The ease with which the
reaction proceeds is directly related to
the property or behavior of these
particular MO's, connecting these to
the phenomena of orientation or
stereoselection. It is understood that
the direct "motive force" which drives
even a complicated organic molecule
to chemical reaction may be ascribed
to merely one electron (or sometimes
more) whose mass is less than a ten or
hundred thousandth of that of the
molecule. (154 references)

Prices are subject to change without
notice

All-Valence Electrons S.C.F. Calculations

By **G. Klopman** and **B. O'Leary**
With 4 figures. 90 pages. 1970
(Topics in Current Chemistry,
Vol. 15/Part 4)
DM 38,–; US $ 14.10

The most important all-valence electron methods proposed for S.C.F.
calculations of the properties of large
organic molecules are discussed. The
last five years have seen the birth of
such methods and the incredibly fast
development of a number of more
efficient variants designed to give
better agreement with specific properties. The trend is undoubtedly in
favour of the development of an "all-
purpose" method. Some authors believe
this involves the development of an
NDDO method. Such a procedure,
however, would require the calculation of a much larger number of integrals and it would be difficult to apply
it to large organic molecules of "chemical interest".
In the opinion of the present authors,
such calculation would not improve
agreement with properties found by
experiment because it would not introduce any fundamentally new feature to make up for the inadequacies of
the present ones. As a matter of fact,
the neglect of two-center integrals
involving one-center differential overlap seems to be a reasonable hypothesis, as shown by the success of the M
(INDO) methods. On the other hand,
research workers have usually confined
themselves to trying to find the best
approximation for molecular integrals,
while overlooking the possibility that
atomic orbitals in molecules might
differ widely from those in the isolated
atoms. (60 references)

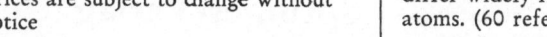

Springer-Verlag
Berlin · Heidelberg · New York
München London Paris Sydney Tokyo Wien

σ and π Electrons in Organic Compounds

By **W. Kutzelnigg, G. Del Re,** and **G. Berthier**
With 11 figures
IV, 122 pages. 1971. (Topics in Current Chemistry, Vol. 22)
DM 48,–; US $ 17.80

In the development of quantum chemistry few concepts have proved to be as significant as the distinction between σ and π electrons in organic compounds. This distinction suggested the approximation known as the 'σ-π separation' which has made it possible to calculate many important physical and chemical properties of unsaturated compounds. (148 references)

Computers in Chemistry

With 13 figures
III, 195 pages. 1973. (Topics in Current Chemistry, Vol. 39)
DM 62,–; US $ 23.00

Introduction

A. J. Thakkar: "The Coming of the Computer Age to Organic Chemistry. Recent Approaches to Systematic Synthesis Analysis." The use of computer programs to assist synthesis will become routine. Four recent approaches to systematic synthesis analysis are reviewed. (23 references)

J. Dugundji and I. Ugi: "An Algebraic Model of Constitutional Chemistry as a Basis for Chemical Computer Programs." A theory is stated which can serve as the foundation of a new type of matrix-fitting computer programs for establishing constitutional relations such as synthetic design, the elucidation of reaction mechanisms, and biosynthetic pathways, as well as the analysis of mass spectra. (37 references)

D. C. Veal: "Computer Techniques for Retrieval of Information from the Chemical Literature." The trend is towards true information retrieval, rather than mere document retrieval. (58 references)

J. T. Clerc and F. Erni: "Identification of Organic Compounds by Computer-Aided Interpretation of Spectra." Applications are in fields where samples have to be routinely analyzed without knowledge of their origin. (42 references)

A. B. Delfino and A. Buchs: "Mass Spectra and Computers." Much emphasis is put on the artificial intelligence approach where the computer is used as a symbol manipulator. (49 references)

F. Caesar: "Computer-Gas Chromatography." The use of computers in gas chromatography will become more and more attractive. (98 references)

W. Geist and P. Ripota: "Computer-Assisted Instruction in Chemistry." Examples are given of programs written to drill students. (47 references)

Prices are subject to change without notice

**Springer-Verlag
Berlin
Heidelberg
New York**
München London Paris
Sydney Tokyo Wien